ESSENTIAL CAR CARE FOR WOMEN

JAMIE LITTLE & DANIELLE MCCORMICK

SEAL PRESS

Essential Car Care for Women
Copyright © 2013 by Jamie Little and Danielle McCormick
Photos of Jamie © Michael Voorhees
Snow chain illustrations © Tim McGrath
Photos on pages 112, 117, 120, 123, and 124 Courtesy of Thule, Inc., www.Thule.com

Published by
Seal Press
A Member of the Perseus Books Group
1700 Fourth Street
Berkeley, California

Library of Congress Cataloging-in-Publication Data
Little, Jamie, 1978–
Essential car care for women / Jamie Little and Danielle McCormick.
p. cm.

ISBN 978-1-58005-436-2 (pbk.)
1. Automobiles—Maintenance and repair—Amateurs' manuals I. McCormick, Danielle,
1978- II. Title.
TL152.L54 2012
629.28'722—dc23
2011045604

9 8 7 6 5 4 3 2 1

Cover and interior design by Kate Basart/Union Pageworks
Printed in China at RR Donnelley Shenzhen
Distributed by Publishers Group West

ESSENTIAL CAR CARE FOR WOMEN

CONTENTS

CHAPTER ③ CAR MAINTENANCE...............................43

CHAPTER 9   SELLING A CAR....................................155

INTRODUCTION

A Word from Jamie

Women today are hands-on, go-get-'em, take-charge ladies. Women have proven they can do anything they set their minds to. Whether we're becoming doctors, engineers, fighter pilots, sports reporters, or simply juggling a career while being wives and/or mothers, we are doing it *All!* This includes taking on the common tasks men are known for doing . . . like taking care of our cars.

Women own 50 percent of the cars on the road today. They play a more significant role in deciding on the family car than men do, yet the stigma remains that we girls just don't know as much about cars as men. The fact is, we are just as smart as boys. It's all just a matter of education, and it's time for a change, ladies! Aside from a house, a car is the second-biggest investment you are likely to make. If you don't take care of it properly, it can end up becoming a money pit, and that can greatly affect the price when it comes time to sell it. Taking control and maintaining your car can save you hundreds, if not thousands, of dollars in a time when everyone is looking for ways to save money.

This book is designed to teach women and girls everything they need to know about owning and maintaining a car. Inside, we explain how a car runs, and we list key terminology. Now, the next time you're watching a NASCAR race or find yourself talking to a mechanic, you'll be educated and prepared, and you will have a better understanding of the automobile.

Essential Car Care for Women will educate you on everything from how to maintain your car to extend its life to how to save money on mechanics.

We give you step-by-step instructions on how to handle car trouble, such as changing a flat tire and what to do when you're involved in an accident. Let's face it: Not everyone has access to AAA or road-side assistance. And when it comes time to sell your car, you will know exactly what to do, thanks to our helpful buying and selling guide.

We hope you find the following pages empowering and helpful—on the road and in your car.

A Word from Danielle

When it comes to cars, the common belief is that women should just leave it to the men. But why should that be? Anyone who went to school with boys will know that we girls are just as capable as they are, if not more so. What's more, the notion of "leaving it to the men" assumes every woman who has a car also has a reliable man in her life.

Your car is a piece of machinery, and as with most machines, there are things you are supposed to do to it along the way to make sure it works properly. You wouldn't get into an aircraft that hadn't been regularly checked by an engineer, would you? Yet every day, millions of people get into cars that haven't had any maintenance in years! By the time smoke is coming out of your engine, you have probably done a lot of expensive damage and significantly reduced the amount of money you will get for the car when you want to sell it.

Knowing what to do to look after your car is not hard. As with anything in life, the information is all there; you just need to know how to find it. *Essential Car Care for Women* is designed to teach a woman everything she needs to know about her set of wheels.

I recommend that you read through the book once so you can learn all about your car and the things you have to do to keep it in prime condition. Then stash the book in your glove box so it will be right there when you need it, and so you don't have to go running to find a man!

Chapter One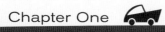

HOW YOUR CAR WORKS

HAVE YOU EVER THOUGHT THAT SOMETHING WAS REALLY HARD and "not for you," and then you had to try your hand at it and realized it was actually quite easy? This will probably be the case when it comes to learning how your car works. If boys can understand it, then there is absolutely no reason why a girl shouldn't be able to!

Understanding how your car actually works is a fascinating journey. Knowing what happens under your hood and how all the components make your car move should mean that the next time you have to take your car to a mechanic, hopefully you will understand what he is talking about.

Once you know how your car works, then understanding car lingo should be no problem!

LET'S GET STARTED

When the automobile was first invented, there was a lot of excitement about this amazing piece of machinery that could transport people long distances in great comfort. However, now cars have become such an ingrained part of our everyday lives that we completely take them for granted. When you are sitting in your car, you are sitting in a highly sophisticated piece of machinery, but you may have no clue as to how anything inside it works!

We tend to care more about things when we understand them, so hopefully when you appreciate how all the different parts of your car are working for you, you will return the favor and give your car the ongoing TLC it deserves.

There are thousands of different makes and models of cars on our roads today. The majority of modern cars are based on the same type of engine, but each model has slight variations in terms of how all the components work together. In order to explain to you how your car works, we are going to describe how a "typical" engine operates.

FROM THE TIME YOU TURN YOUR KEY IN THE IGNITION TO THE moment your wheels actually start turning, there are lots of different processes happening at great speeds. Before we go through a description of the process from start to finish, there is one place that is worth looking at in more detail, because that's where all the action takes place—the cylinders.

Some wonderful scientists discovered that if you add a tiny amount of high-energy fuel (such as gasoline) to a larger volume of air (roughly one part gasoline to fourteen parts air), you can create a very useful gas. When this gas is put in a small enclosed space and ignited, it releases an incredible amount of energy. Then some exceptionally clever scientists discovered that they could harness this energy in a useful way to get machines to carry us around. The next thing you know, the noble horse was out of a job and the automobile came into being.

Almost all of today's modern car engines use what is known as a four-stroke combustion cycle to create movement. To put it in simpler terms, there are four movements that happen in a particular order. These movements consume fuel and result in the release of a useful energy.

The main components you need to understand inside the cylinder are:

[1] The PISTON—A sliding piece of metal that moves up and down inside the cylinder.

[2] The INTAKE VALVE—The part of the cylinder that allows the air-fuel mix into the chamber.

[3] The EXHAUST VALVE—The part of the cylinder that releases the exhaust (used) fumes from the chamber.

[4] INTERNAL COMBUSTION CHAMBER—The enclosed place, created by the pistons moving down the cylinder, that contains the explosions made by ignited fuel.

[S] The SPARK PLUG—Produces the spark that ignites the air-fuel mix.

Four-Stroke Combustion Cycle

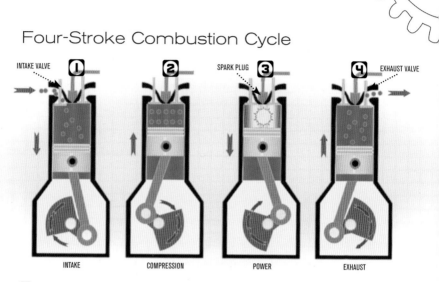

INTAKE VALVE ① | ② | SPARK PLUG ③ | ④ EXHAUST VALVE

INTAKE | COMPRESSION | POWER | EXHAUST

① The first stroke is known as the intake, or induction, stroke: The cycle begins with the piston at the top of the cylinder. The intake valve opens and the piston moves down the cylinder, allowing the air-fuel mix to enter the open space.

② Next is the compression stroke: The piston now moves back up toward the top of the cylinder, thereby squeezing the air-fuel mix into a smaller space (i.e., compression). This will ultimately make the explosion more powerful.

③ When the piston has reached the highest point in the cylinder, the spark plug emits a spark, which ignites the air-fuel mix and causes an explosion. The power of this explosion pushes the piston back down the cylinder with great force. The bottoms of the pistons are attached at right angles to a shaft with offset bearings, called a crankshaft. The up-and-down motions of the pistons in this cycle rotate the crankshaft in a way similar to the up-and-down motion of your legs on a bicycle. Ultimately, the motion that comes out of the engine is rotational, which, funnily enough, is exactly the motion needed to turn the wheels of a car! This rotational force is known as torque.

④ Finally, after the explosion, the exhaust valve opens and the piston moves back up to the top of the cylinder, forcing the exhaust fumes out of the exhaust valve. Now that it is at the top of the cylinder, it is ready to start the cycle again.

Revolutions per Minute (RPM)

In a car, this cycle repeats itself thousands of times per minute inside each cylinder. This is where the expression "revs (or revolutions) per minute" comes from. Your rev counter is telling you how many thousand times per minute these explosions are turning the crankshaft. Next time you are driving, check out your rev counter and see how hard your engine is working for you!

Cylinder Banks

V-SHAPED CYLINDER BANK

There are a number of cylinders in your engine. The number you have depends on what type of car you have. The cylinders can be arranged in various different shapes: "In-line" means they are arranged in a row; "V" means they are arranged in two banks (rows) in a "V" shape; "flat" means there are two opposing rows. When someone says they have a "V6" or "V8" engine, it means there are six or eight cylinders arranged in a "V" shape.

AND THERE YOU HAVE IT. NOW THAT YOU HAVE A BASIC UNDERSTANDING OF HOW ENERGY IS CREATED TO MOVE THE VEHICLE, LET'S LOOK AT THE PROCESS FULLY SO YOU'LL HAVE A BETTER UNDERSTANDING OF HOW ALL THE OTHER PARTS OF YOUR ENGINE ARE WORKING TOGETHER TO MAKE YOUR CAR MOVE.

What Happens When I Turn the Key in the Ignition?

When you turn the key in the ignition, the **battery** powers the **starter motor**, which begins to turn the crankshaft to get the pistons moving.

Air then enters your engine via a filter that removes any dirt or grit from the incoming air. At this point, fuel (either gasoline or diesel) is added to the air to create a vaporized gas. This gas is now waiting in a chamber for you to decide how much of it goes into the engine. The amount of gas going into your engine is controlled by your foot on the accelerator pedal. If you want to add more fuel to your engine (to go fast or to climb a hill), you must press the accelerator pedal down farther. This opens the **throttle valves** wider, allowing a larger amount of gas through. If you only need a little bit of power (and therefore a little bit of fuel), you put your foot down lightly on the pedal, which will open the throttle valve just a fraction.

From here, the gas goes through what is called an **intake manifold**, which essentially distributes the gas to each of the cylinders through a series of passages and **valves**. The opening and closing of the valves is carried out by the **camshaft**.

In most cars, when air enters the engine (to be mixed with fuel), the way it comes in is the way it is used. In some cars, however, the air is pressurized so more air-fuel mix can be squeezed into the cylinder to increase performance. This is referred to as **"turbocharged."**

Now We Come to the Cylinder Bit

The gas then enters each cylinder via the intake valve. The piston comes up to the top of the cylinder, and at this point, the valves are closed.

At the exact same time, the distributor causes a spark plug to spark, which ignites the fuel in the cylinder and causes an explosion. The explosions that occur inside the different cylinders are timed to go off at different intervals, ensuring the crankshaft is continually spinning. The force of the explosion pushes the piston down the cylinder sharply. The pistons are attached at a right angle to the crankshaft. As the piston is forced down, it causes the crankshaft to rotate.

So now your pistons are making the crankshaft rotate rapidly. But before all this motion goes to your wheels, you need to be able to control

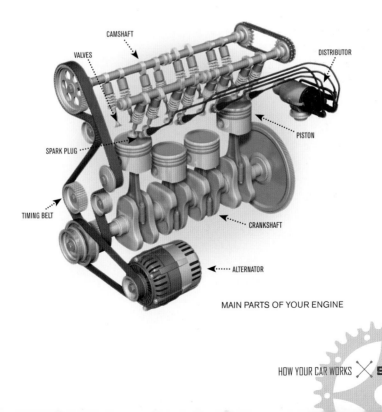

CAMSHAFT

VALVES

DISTRIBUTOR

PISTON

SPARK PLUG

TIMING BELT

CRANKSHAFT

ALTERNATOR

MAIN PARTS OF YOUR ENGINE

it (so your car doesn't go one hundred miles per hour when you only need to go thirty). Therefore, the crankshaft goes through a selection called the transmission. This section of the car is in charge of controlling the power contained in the crankshaft before it goes to the wheels.

The transmission controls the speed/power of your car by providing you with different speed/power ratios otherwise known as gears. For example, in first gear, you need a lot of power to initially get your car moving, but you don't need a lot of

speed. In a higher gear, the same amount of power would get you to a higher speed. The transmission is responsible for regulating all of this. If you have a manual car, you control it with your stick shift.

The transmission is connected to the output shaft, which is connected to the axles, which are in turn connected to the wheels. When the transmission turns, the output shaft turns the axles, and that turning then rotates your wheels. *Et voilà*—you're moving! Now that you know how all the main engine parts work to make your car move, it's time to look at the other vital components.

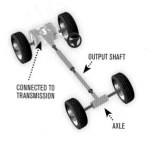

The Electronic Control Unit (ECU)

In recent years, cars have become much more high-tech and now even have computers in them that program certain elements of your engine. The electronic control unit controls lots of things, such as the fuel injection system, air-conditioning, air flow, and idling speeds (i.e., when you are sitting in traffic). In some instances, when you bring your car to your mechanic or dealership, they will be able to plug a laptop into your car and get readings of how all the different components attached to the system are working.

The Oil

There are lots of moving parts in your engine, and oil makes sure that all the parts are lubricated so they can move easily.

Usually oil is sucked out of the oil pan by a pump, which then passes it through a filter to remove any dirt, and then squirts it under high pressure onto the bearings and the cylinder walls. The oil then trickles into an area called the sump, where it is collected. The process then starts over again.

The main parts that need oil are:

1 The PISTONS—So they can slide up and down the cylinders.

2 The CAMSHAFT and the CRANKSHAFT—They have bearings that enable them to move freely, and oil is used on these bearings to help them move.

The Alternator

The alternator is an important player in the electrical system of your car.

When the battery has started your car engine, a belt within your engine begins to rotate. The movement of this belt drives a device known as the alternator. The alternator produces electricity by converting mechanical energy into electrical energy.

This electrical energy is used to power all the electrical components in your car, such as the ignition lights, heater, wipers, radio, etc. It also puts back into the battery the energy that the start motor has used.

The Regulator

This is a device that regulates the amount of energy in the alternator to ensure it has the exact amount of energy it needs.

The Distributor

The distributor gets the ignition coil to generate a spark at the precise instant at which it is needed. It is also responsible for directing (i.e., distributing) the spark to the right cylinder at the right time. It has to do this at exactly the right instant and up to several thousand times a min-

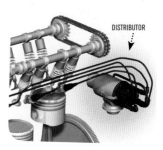

DISTRIBUTOR

ute for each cylinder in the engine. If the timing spark is off by even a fraction of a second, the engine will run poorly or not at all.

The distributor is now being replaced in more modern engines with distributorless ignition systems or separate coil packs. These provide the same function as the distributor, except they are electronically con-trolled by sensors.

The Timing Belt

TIMING BELT

The camshaft (at the top of the cylinder) and the crankshaft (at the bottom of the cylinder) need to work synchronously. The timing belt is a belt that connects the two, ensuring that they work in time with each other.

Consult your owner's manual to find out how often your timing belt needs to be changed.

DEPENDING ON THE MODEL OF YOUR CAR, TIMING BELTS USUALLY NEED TO BE REPLACED EVERY 60,000–100,000 MILES.

The Cooling System

With all the fuel being burned, as you can imagine, the engine will get quite hot! As a result, your car has a cooling system, which keeps the temperature of the engine down. The cool-ing system consists mainly

CAR RADIATOR

of the radiator, water pump, and temperature gauge. Water circulates through passages around the cylinders and then travels through the radiators to cool down.

The Exhaust System

Remember when the gas was burned in the combustion chamber? As soon as it has been burned, it exits the combustion chamber via an exhaust valve and enters an exhaust system.

Most modern cars have a catalytic converter in their exhaust system that burns off any unused fuel and certain chemicals before it is released from the car via the exhaust pipe. The catalytic converter minimizes the amount of toxic fumes coming from your car.

What Is the Difference Between a Diesel and a Gasoline Engine?

Diesel and gasoline cars are basically the same, but the main difference between the two is how the explosions that take place in the cylinders occur. In a gasoline engine, fuel is mixed with air, and then it is forced into the cylinders, where this air-fuel mix is compressed by the pistons and ignited by sparks from spark plugs.

However, in a diesel engine, the air is compressed before the fuel is added to it. When air is compressed, it heats up. So, by the time the fuel is added to the compressed air, it is very hot, and the air-fuel mix ignites automatically. There is no need for a spark plug to create a spark. In short, in a gasoline engine, a spark plug ignites the explosion in the cylinders, whereas in a diesel engine, pressure causes the ignition.

The Head Gasket

The head gasket is tucked away inside your engine, so it's not something that you see when you open up your hood, but if it blows, you can expect a big bill from your mechanic!

What Is It and Why Do I Need to Worry?

The **cylinder head** (the block that seals all the tops of your cylinders) is made in one part of the car factory, and the **engine block** (which contains all the main bodies of the cylinders) is made in another part of the factory. When it comes to putting your engine together, these two pieces need to fit together seamlessly. With all the explosions taking place inside the cylinders, there is no room for cracks or open spaces! So, in order to ensure the cylinder head and engine block fit together seamlessly, the car people put a piece of metal called a head gasket in between them.

If your engine overheats for a sustained amount of time, the head gasket can warp or crack and eventually blow. Replacing a head gasket is very labor-intensive and therefore very costly. That is why it is really important that as soon as your engine shows any sign of overheating, you should pull over as soon as it is safe to do so and call for roadside assistance. It is generally much cheaper to call for roadside assistance than to replace a head gasket.

KEEPING A CONSTANT WATCH ON YOUR OIL AND COOLANT LEVELS WILL HELP PREVENT THESE SITUATIONS FROM HAPPENING. THIS IS WHY OUR DADS|BOYFRIENDS|BROTHERS KEEP NAGGING US TO DO THESE SEEMINGLY UNIMPORTANT LITTLE THINGS. NOW WE KNOW WHY THERE IS ACTUALLY A GOOD REASON WHY WE SHOULD LISTEN TO THEM!

Filters

There are lots of moving liquids and gases in your car. Filters are set in strategic places along each system to catch and trap anything that might contaminate it. Your car has several types of filters, such as an oil filter, an air filter, and a fuel filter.

These need to be changed quite regularly, and your mechanic should do this for you when you service your car.

Belts and Hoses

Your car has various belts, which are used for rotating different components in your engine (e.g., fan belt, alternator belt).

Hoses transport fluid or gases from one part of the engine to another. If a belt cracks or a hose gets a hole in it, it can severely affect the performance of your engine.

NOW THAT YOU KNOW WHAT ALL THE DIFFERENT PARTS OF YOUR ENGINE
DO, LET'S SEE WHERE THEY ARE WHEN YOU OPEN YOUR HOOD. NOT ALL CAR
ENGINES ARE THE SAME, SO BE AWARE THAT THE PARTS OF YOUR ENGINE
MIGHT BE ARRANGED SLIGHTLY DIFFERENTLY.

1 Intake Manifold

2 Cylinders

3 Electronic Control Unit
(not visible)

4 Spark Plugs
(inside the cylinders)

5 Radiator

6 Battery

7 Air Filter
(inside the housing)

Brakes

There are usually two different types of brakes in your car: disc brakes and drum brakes. Disc brakes work in a very similar way as the brakes on a bicycle. On a bicycle, there is a piece called a caliper that squeezes the brake pads against the wheel. In a car, the disc brakes have a caliper (guarded with brake pads) that grabs onto the disc of the wheel to get it to stop. Drum brakes work on the same principles as disc brakes. However, a drum brake presses against the inside of the drum.

TYPICAL DISC BRAKE SYSTEM

The force used to get the brake pads to work is transmitted hydraulically, or through a fluid. When you step on your brake pedal, you are actually pushing against a plunger that forces brake fluid through a series of tubes and hoses. This ultimately puts pressure on the brake pads to stop your wheels.

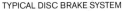

THE BRAKE PADS WEAR DOWN OVER TIME AND NEED TO BE REPLACED QUITE REGULARLY. IF YOU HEAR NOISES WHEN YOUR CAR IS BRAKING, IT'S TIME TO TAKE IT TO YOUR MECHANIC!

Suspension System

Have you ever bounced a ball on the ground and landed it on a stone? You may remember that the stone caused the ball to go in a completely different direction than the way you wanted it to go. Similarly, without a suspension system in your car, every time your tires hit a stone or bump, it would send the tires off in the wrong direction. This would make the car difficult to steer and uncomfortable for any passengers. For this reason, your car has a suspension system that has been designed to counteract the effects of your car's hitting imperfections on the road.

The suspension system is made up of springs and shock absorbers. The main body of the car is attached to the base of the car by springs. These springs absorb any of the bouncing energy so the frame and the car remain undisturbed as you drive.

Shock Absorbers

If the spring in the suspension system were left to its own devices, it would bounce up and down until it ran out of energy, which wouldn't be very pleasant for anybody sitting in the

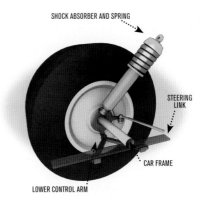

SHOCK ABSORBER AND SPRING

STEERING LINK

CAR FRAME

LOWER CONTROL ARM

car. A shock absorber is a device that absorbs the energy from the springs to stop the bouncing motion.

Tires

As the part of the car that actually connects with the road, the tire plays a vital, but often overlooked, role.

Tires are made of rubber and have a tube in the middle of them. If the tube gets damaged, it deflates and you have a flat tire! The rubber and tubing create a soft ride for you and are also part of your suspension system.

In order to grip the surface of the ground, tires have a tread cut into their surface. With all the friction caused by rotating at great speeds along the road, the tread in the tire wears down. If the tread becomes too thin, it inhibits the tire's ability to grip wet roads and can cause accidents. This is why in most countries it is now a legal requirement that the tread of a tire not get to less than 1.6 millimeters.

Tires and Wheels

A wheel consists of the metal rims that support the rubber tires. There are a few different types of tires and wheels available on the market today.

Standard Wheels

The most common wheels have a thick rubber tire with a steel rim (the piece of metal that supports the rubber tire) and a hubcap (a wheel cover that fits over the hub of the wheel).

Alloy Wheels

Alloy wheels are made of an alloy (two metals mixed together) based on either aluminum or magnesium. They tend to be lighter than the standard wheels, which helps improve steering and speed.

RUBBER TIRE

TIRE TREAD

STANDARD TIRE

Low-Profile Tires

Low-profile tires have a very thin tire base. This lessens the distance between the main body of the car and the ground, which makes them look sportier and gives them a better grip on the ground—although with less air in the tires, you will feel every bump!

These tires tend to be more expensive and don't last as long as standard tires, but they help the car handle the roads better at higher speeds, so they are a favorite among speed demons.

Run-Flat Tires

Some luxury and sports cars are now coming with run-flat tires. Run-flats contain a sensor that detects any loss of pressure and relays this message to the dashboard in the event of a flat. They have reinforcements built into them so that you will have enough time to drive to the nearest garage without having to change the tire yourself. The downside of run-flats is that they have to be replaced with another run-flat tire, and they are more expensive than standard tires.

ALLOY WHEELS

LOW-PROFILE TIRES

RUN-FLATS

[NOTE: FOR MORE INFORMATION ON SNOW TIRES, REFER TO CHAPTER 7, DRIVING IN THE SNOW.]

Chapter Two

GETTING DOWN WITH THE LINGO

WHEN PEOPLE ARE TALKING ABOUT CARS, SOMETIMES IT CAN FEEL as if they are speaking another language. They talk excitedly to you about their new "three-liter V6 TDI coupe with 240 horsepower" like you know exactly what they mean, when the truth is that you haven't got a clue. After reading this section, though, you should be able to talk with confidence to even the biggest car enthusiast.

PERFORMANCE CHARACTERISTICS

What, Exactly, Does "Liter Size" Mean?

If you can remember back to your days in science class, cylinders typically held liquids, and the units we use to measure liquids are liters. Similarly, in the cylinders of your car, the space that is available for the air-fuel mix to fill is also measured in liters (or cubic inches, in older muscle cars). The more space you have available, the more air-fuel mix you can get in there, which ultimately means more power.

One cylinder might have 0.5 liters of space available for the air-fuel mix. But remember, your car usually has a number of cylinders, so if you had six cylinders with 0.5 liters of space available in each, 6 x .05 = 3 liters. This measurement is sometimes called the engine displacement, or volume.

A very small, economy-size car might have only small cylinders, which in total might amount to 1 liter, whereas a high-performance car's cylinders might add up to 6 liters, like a Ferrari's.

CYLINDER

Having a large-size engine might sound impressive, but the bigger the size of the engine, the more fuel you will need to run it. The operating costs of cars with big engines can be very expensive!

Valves

Each cylinder in the engine has valves; valves allow the air-fuel mix into the cylinder and allow the exhaust fumes out. The minimum a cylinder requires is two valves—one for each function. Some car engineers design the cylinders to have more than the standard two valves, as  this enables more air-fuel mix to enter the cylinders. If a car has four cylinders with four valves in each, you would say it has a "four-cylinder, sixteen-valve engine." Generally, more valves per cylinder means higher performance in terms of horsepower, even for the same engine volume.

Horsepower (HP)

There is no doubt that even the biggest car enthusiast would struggle to have a firm grip on horsepower and its exact definition, but it simply means the amount of power it takes to move thirty-three thousand pounds the distance of one foot in one minute. This may seem like an odd measurement, but it dates back to the time when people began transitioning from horses to cars.

History of Horsepower

Just think back to an old movie set in a time before cars existed—horses were used a lot in heavy lifting and in transporting people around. So, back in the day, when famous scientist James Watt (yes, the same guy whose name is written on your lightbulb!) was trying to convince people of the amazing powers of the new steam engine, the best way he could do that was by comparing the power of the new engine to the power of a horse.

James Watt came up with a calculation that one horse could do thirty thousand foot-pounds of work in one minute. This is not exactly a straightforward measurement, but for some reason, it has stuck as a measurement for calculating the power of engines—including cars.

Unless you are a car engineer, there is no need to be a pro at horsepower. For most people, understanding that horsepower is a way of measuring the power of a car and knowing what is relatively good and what is relatively poor is enough.

The table below highlights the horsepower of some well-known cars that are good benchmarks for comparing your car against.

You will notice in the chart that the weight of the car is included. This is because if you had two cars with the same horsepower but one weighed twice as much as the other, the heavier car wouldn't be as fast as the lighter one. When race car manufacturers build vehicles to compete against each other for speed, they try to get the highest horsepower they can and build the cars as light as they can. That is why racing cars have only one seat—everything is geared toward keeping the weight of the car down.

	CAR	WEIGHT	HORSEPOWER
	FORD FOCUS	2.470 LBS	136
	PORSCHE	2.975 LBS	300
	FERRARI ENZO	3,009 LBS	650
	FORMULA 1 CAR	1,323 LBS	900

Brake Horsepower (BHP)

Brake horsepower is the measurement of the car's horsepower when it comes straight out of the crankshaft. It loses some of its power when it goes through the gearbox and other components, so the actual horsepower that is delivered to the wheels tends to be a lower figure than the bhp, which is produced from the engine.

Torque

The energy that moves through your car is rotational, and torque is a measurement used to measure a force that occurs in a twisting motion. It is measured in foot-pounds (ft-lbs) and is another way of measuring the power of the car.

Zero to Sixty

Sometimes in car talk you hear the expression "It does zero to sixty" in X number of seconds. Cars come in all shapes and sizes and with different engines, so it can be hard to compare one against the other. The "zero to sixty" method has become a way of comparing the acceleration of two different cars, by stating how many seconds it takes one car to get from a starting position (i.e., zero miles per hour) to sixty miles per hour. For example, a Ford Ka can get from zero to sixty in 11.8 seconds, whereas a Ferrari can go from zero to sixty in 3.3 seconds.

In the metric system, this range is measured from zero to one hundred kilometers per hour.

Miles per Gallon

Miles per gallon refers to how far you will go on one gallon of gasoline. The more miles your car will do for a gallon of gasoline, the better it will be for your pocketbook.

TYPE OF CAR	MILES PER GALLON
LAMBORGHINI	8
TOYOTA COROLLA	30
TOYOTA PRIUS	45

As you will see from the chart above, having a fancy sports car may be all very well and good, but your gasoline costs will be very expensive. On the other hand, the new hybrid cars should cost very little to run. In a year, if a Lamborghini were to do the same amount of miles as a hybrid car, the Lamborghini owner's gasoline bill would be nearly six times as much as the hybrid owner's. It gives you something to think about the next time you are buying a car!

In the metric system, miles per gallon translates to kilometers per liter.

ENGINE FEATURES

Turbocharger

In cars, ideally you want to try to get as much power out of your engine as you can. One way of doing this is by packing as much of the air-fuel mix into the cylinders of the engine as possible. A turbocharger is a device that uses the pressure from the exhaust to create more pressure in the cylinders so that they can receive more of the air-fuel mix.

Turbo

If a car is described as being "turbo," that means it has a turbocharger in its engine.

Twin Turbo

"Twin turbo" means that the car has two turbochargers. Each turbocharger is driven by one-half of the exhaust system. If the car has two exhaust pipes, then a turbocharger would be connected to each one.

Supercharger

A supercharger performs the same function as the turbocharger inasmuch as it increases the amount of air-fuel mix the cylinders can take. While the turbocharger is run from the pressure of the exhaust system, a supercharger gets its power from a belt or is sometimes attached to the engine's crankshaft. A supercharger is something that can be added to a car after you have purchased it and is a way of increasing power to the engine.

Carburetor

The carburetor is a device that mixes the fuel and the air before it is passed on to the engine for combustion. It is a simple, noncomputerized device, but it has generally been replaced in cars since the 1980s by the more efficient fuel-injection system.

Fuel-Injection System

More modern cars now have a computerized system that mixes the air and fuel before it goes into the engine. This is called a fuel-injection system. Because it is computerized, it is considered more efficient than its predecessor, the carburetor, and makes sure the correct mix is put in every time.

TDI

TDI stands for turbo direct injection.

TDI is something that is found only in a diesel engine. It used to be the norm that in diesel cars, the air-fuel mix would have to go to a pre-combustion chamber before it was sent to the main cylinders of the engine. However, in a TDI car, the fuel is injected directly.

A TDI car also has a turbocharger. With both direct injection and a turbocharger, it makes for a more powerful car.

GTi

GTi is an expression coined by car engineers. It stands for "gran tourismo injection" or "grand tourisme injection." It comes from the phrase *grand tourer,* which has become synonymous with high-powered sports cars designed to race long distances. The "injection" part means that it has fuel injection, as opposed to a carburetor.

DRIVING CHARACTERISTICS

Car Handling

Car handling refers to how responsive the car is to your steering. If you are trying to drive around a corner and your car seems to go a different direction than the way you want it to, then you would say the car doesn't have good handling. If you are driving around a corner at a very fast speed and your car does exactly what you want it to, then you would say the car has great handling. It can also refer to when you are driving in a straight line. If you are keeping your steering wheel straight and your car keeps swerving in a different direction, then this is a sign of poor car handling or poor alignment.

Understeer

If you are approaching a corner and the steering wheel doesn't turn the wheels as much as you want it to, this is called "understeer." This is also sometimes referred to as "tight" or "push."

Oversteer

If you are turning a corner and the rear wheels do not follow the front wheels and instead veer toward the outside of the turn, this is known as "oversteer." This can sometimes lead to skidding, which is quite dangerous. This is also known as "loose" or "free."

Traction

This refers to how well the tires grip the ground. New tires with deep tread tend to have good traction. Old tires that are worn and have very little tread left have poor traction on a wet surface. On the other hand, bald tires work best on a dry surface; if you ever see a drag racer's tires, they are completely bald for this reason.

Rear Wheel Drive (RWD)

In a rear wheel drive car, the power created by the engine is distributed to the rear wheels, and these in turn drive the car. RWD tends to be a feature found in high-performance cars because it makes the car handle better in dry conditions. This means drivers can turn corners at greater speeds than they can in a normal car. It also offers better braking ability.

RWD is more expensive to put in place, so a lot of mass-produced cars have front wheel drive.

Front Wheel Drive

When the engine drives only the front wheels of the car.

Four Wheel Drive

Also known as 4 x 4. This describes when the engine powers all four wheels.

ABS

"Anti-lock brakes" or "anti-locking brake system." When you have to brake suddenly, it is usually because there is an emergency and it is the moment you need your brakes the most. Some brake systems can sometimes "lock" at the moment of greatest need, so car manufacturers developed a way of stopping the brakes from locking in these emergency situations so that you can have more control.

Power Steering

It used to take a lot more effort to turn the wheels of the car with the steering wheel before power steering was invented. If a car has a power steering system, it means that there is a separate power source that is helping to turn the wheels.

Automatic

If a car is described as an "automatic," it means that it has an automatic transmission or gearbox. The transmission will automatically shift the gears as it accelerates and decelerates, so the driver does not need to keep on changing the gears manually. Typically, the functions of an automatic gear shift are "park," "drive," "reverse," and "neutral."

Manual

A manual car is a car in which the driver physically changes the gears up and down. A manual transmission allows drivers to decide when they want the car to go into a higher or lower gear, and therefore gives them more control.

TYPES OF CARS

Sports Car

There is a lot of debate among car enthusiasts as to the correct definition of a "sports car." The hardcore enthusiasts would say that a sports car must be a car that has been built solely for performance, rather than practicality. These cars tend to be built for speed, rather than for family use, so they have only two seats, are quite low to the ground, and have aerodynamic bodies. These are usually made by sports car specialists, such as Ferrari, Porsche, and Lamborghini.

High-Performance Car

Mass-market manufacturers, such as BMW, Mercedes, and Ford, have begun to come out with high-performance versions of their existing models, which have some of the features of a sports car, such as a more powerful engine or a sleeker look. Some people consider these to be sports cars, but die-hard enthusiasts would just call them "high-performance cars."

Hybrid Car

A hybrid car has both a gasoline and an electric engine. A traditional gasoline car can waste a lot of fuel, as it needlessly burns through fuel while you are sitting in traffic or going downhill while the car has natural momentum. The hybrid has sensors in it that automatically shut down the gasoline engine whenever it doesn't need to be used, such as when you are sitting in traffic, and switch to the electric engine. Conversely, since an electric engine doesn't have enough power to drive

CAR BODY STYLES

A SALOON, OR SEDAN, IS A CAR THAT HAS TWO FRONT SEATS AND AT LEAST TWO REAR SEATS, WITH FOUR DOORS AND A SEPARATE TRUNK.

A COUPE IS OFTEN THE SPORTY VERSION OF A SALOON CAR. COUPES USUALLY HAVE ONLY TWO DOORS, AS OPPOSED TO FOUR. THE BACK SEATS CAN BE QUITE CLOSE TO THE FRONT SEATS.

AN ESTATE IS ALSO KNOWN AS A STATION WAGON. IT HAS AN EXTENDED REAR CARGO, AND THE FULL HEIGHT OF THE CAR EXTENDS TO THE TRUNK, ALLOWING FOR MUCH MORE STORAGE SPACE IN THE TRUNK.

A HATCHBACK IS LIKE AN ESTATE, BUT IT USUALLY HAS A MUCH SMALLER TRUNK SPACE. THE TRUNK CAN BE ACCESSED FROM THE BACK PASSENGER SEATS.

AN SUV IS ALSO KNOWN AS A SPORTS UTILITY VEHICLE, AN OFF-ROAD VEHICLE, A FOUR WHEEL DRIVE, OR A 4 X 4. AN SUV TYPICALLY HAS TOWING CAPABILITIES, CAN GO OFF-ROAD AND CAN ALSO CARRY MANY PASSENGERS. THERE IS NORMALLY NO DESIGNATED TRUNK SPACE, AS THE TRUNK IS DIRECTLY BEHIND THE SEATS.

MPV STANDS FOR MULTI-PURPOSE VEHICLE. IT CAN REFER TO A MINIVAN OR A PEOPLE CARRIER. IT IS SIMILAR IN SHAPE TO A VAN BUT HAS BEEN DESIGNED FOR PERSONAL USE, USUALLY FOR CARRYING LARGER NUMBERS OF PEOPLE THAN A NORMAL CAR.

A CONVERTIBLE IS ALSO KNOWN AS A CABRIOLET. IT IS ANY CAR WITH A FOLDABLE OR RETRACTABLE ROOF.

your car long distances on a freeway, in these instances a hybrid car switches to the gasoline engine. Hybrid cars are very fuel-efficient and are becoming more sophisticated and advanced every day. Because of ever-increasing gas prices, these are becoming more popular, as they have much lower running costs than a regular car.

Electric Car

An electric car is driven by an electric motor and batteries, as opposed to a four-stroke combustion cycle engine. As there is no combustion in the car, it doesn't produce any emissions directly. However, the batteries need to be charged, and if they are charged from an outlet that gets its power from a CO_2-creating source, then the car indirectly creates emissions. If the batteries are charged from a renewable source such as a wind farm, then you can drive a car safe in the knowledge that you are not creating any harmful CO_2 emissions. Electric cars used to be small and could travel only short distances. However, the technology is improving every day, and now the newer versions can travel up to three hundred miles on one charge, and some models, such as the Tesla, are competing with sports cars for style and elegance.

Biodiesel

Biodiesel cars use alternative fuels, such as oils made from vegetables or animal fats, to run diesel engines. The crops that the alternative fuels are made from are renewable. While growing, plants absorb carbon dioxide (CO_2) from the atmosphere and release oxygen. The CO_2 emissions produced by biodiesel cars are said to be carbon neutral, as any CO_2 produced is soaked up by the crops that are currently growing.

Chapter Three

CAR MAINTENANCE

IT IS MORE THAN LIKELY THAT YOUR CAR IS ONE OF THE MOST expensive purchases you will ever make, so doesn't it make sense to look after it properly?

If you don't look after your car, you can expect more expensive trips to the mechanic—usually when you can least afford it—and you are more likely to get a lower resale value on your car.

Basic car maintenance is actually quite simple when you know how to do it. Get into the habit of setting aside a few minutes every two weeks or so, and carry out the basic checks in this section.

BEFORE WE BEGIN...
OWNER'S MANUAL

Sure, your owner's manual may not be a summer read on the beach, but it is an essential tool to running and maintaining your car. The good news is, you are not expected to read the manual cover to cover—just refer to the relevant sections when you need them.

You may even find that when you do start reading the section on, for example, oil, it is surprisingly easy to understand.

Make sure you keep the owner's manual in a safe place in your car. It is usually designed to fit neatly into your glove compartment, so try to keep it there! It should give you all the basic information, from how to open your hood to how to change your tire. Just look in the index to get the page reference for whatever you need to know.

If your car has lost its owner's manual, try to get another one for your car's precise make and model. You can get them from your car dealer, and many are now available to order online.

SYMBOLS ON YOUR DASHBOARD

Cars are becoming much more sophisticated, and in newer cars, there tend to be a lot more sensors in various parts relaying information to your dashboard to notify you if something isn't right. You should familiarize yourself with the meaning of your dashboard symbols by consulting your owner's manual. Older cars won't be as good at telling you if something is wrong, so you really have to look after them carefully!

AN ORANGE SYMBOL USUALLY MEANS YOU HAVE A PROBLEM AND SHOULD PROBABLY GET IT CHECKED AT YOUR EARLIEST CONVENIENCE. A RED SYMBOL MEANS YOU SHOULD PROBABLY STOP USING THE CAR AND GET IT CHECKED BY A MECHANIC IMMEDIATELY.

Here Are Some Examples of Common Symbols:

 ENGINE WARNING—Depending on your make of car, this usually means there is a problem somewhere in your engine's management system.

 OIL—If this light shows, check your oil levels immediately. A lack of oil can lead to overheating, which can seriously damage your engine.

 BATTERY—If this light shows, it means you are experiencing problems with your battery. It can be one of a number of things:

> Your battery's terminals may need cleaning
>
> Your battery is not being charged properly by the alternator
>
> Your battery usually has a life of three and a half years, so it may be nearing the end of its life and need replacing

 BRAKES—This indicates that you have a problem somewhere in your brake system. You can check your brake fluids to see if they are low, but generally all matters that relate to your brake system should be left to the professionals, so bring your car to a mechanic as soon as possible.

 ENGINE OVERHEATING—This symbol means your engine is overheating. Pull over as soon as it is safe to do so, and call for roadside assistance.

 ABS—If this light appears on your dashboard, it means there is a fault somewhere in your anti-lock braking system and it should be checked by a mechanic as soon as possible.

These are just some of the common symbols on your dashboard. Your car manufacturer may have some more symbols that communicate problems to you.

MAINTENANCE CHECKS

Before You Open Your Hood:

❶ All under-hood checks should be done when the engine is turned off.

❷ If the engine has been running for any length of time, it is likely there will be areas that are very hot!

❸ Read your owner's manual's safety warnings to see if there are any particular safety stickers you should keep an eye out for.

❹ If the engine is running, don't put your hands near any fans or belts.

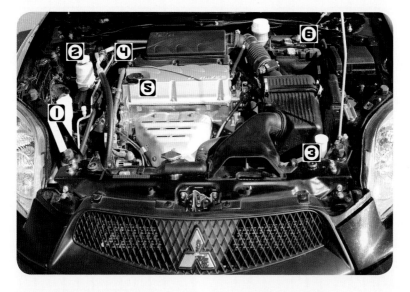

❶ Screen Wash   **❹** Dipstick

❷ Brake Fluid   **❺** Oil Reserve

❸ Coolant   **❻** Battery

CHECKING YOUR OIL

Your car should be stationary, with the engine turned off for at least five minutes, before you check your oil, so you can get an accurate oil reading.

You refill the oil via the oil reserve tank. It usually has a picture of an oil canister on it, but if you are unsure where it is, consult your owner's manual.

There are two different types of oil for cars—synthetic and non-synthetic. Again, consult your owner's manual to see what type your car takes. Most gas stations and garages will stock both types.

Just because you buy a whole container of oil doesn't mean you have to put it all in. Stop every now and then when you are refilling and check your oil levels again to make sure you haven't passed the full mark.

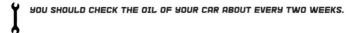

YOU SHOULD CHECK THE OIL OF YOUR CAR ABOUT EVERY TWO WEEKS.

1 LOCATE YOUR DIPSTICK

2 REMOVE THE DIPSTICK

3 CLEAN IT WITH A RAG OR PAPER TOWEL

4 PUT THE DIPSTICK BACK IN FULLY FOR ABOUT 5 SECONDS

5 REMOVE THE DIPSTICK AGAIN SLOWLY

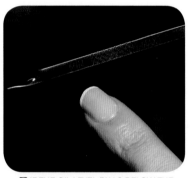

6 IF THE OIL LEVEL FALLS BELOW THE MIN MARK YOU NEED TO REFILL

7 REMOVE OIL CAP

8 ADD OIL

COOLANT | ANTIFREEZE

The cooling system of the car is under high pressure and contains fluid that can heat to a higher temperature than boiling water. Therefore, never open or go near the coolant reserve tank or radiator of a car that has just been running.

Check your owner's manual to find out where your coolant reserve tank is located.

It is usually translucent white, so you can inspect the fluid levels without opening it. Be careful not to confuse it with the wiper fluid tank, as they look similar.

The reserve tank will have marks on the side of it with FULL HOT or FULL COLD. Remember from science class that when things are hot they expand, and when they are cold they contract—hence there are two levels. If your level is below FULL COLD when your engine is cold, you will need to add fluid until it is near that mark.

Check your owner's manual to determine the exact fluid you need to add to your cooling system.

BRAKE FLUID

The brake fluid reservoir is under the hood, usually in front of the steering wheel.

Most cars today have a transparent reservoir so you can see the level without opening the cover.

Changing brake fluid is something that should be left to the professionals, so in this instance, just make sure the brake fluid levels are above the minimum mark.

If you notice a drop in the levels below the minimum, you will need to take your car to a mechanic, as it is an indication that there is a fault somewhere in your braking system.

If the level drops noticeably over a short period of time or goes down to only about two-thirds full, have your brakes checked as soon as possible.

WIPER FLUID

This may not seem very important, but if a big truck has ever splashed mud on your windshield on a wet day, you will know the importance of having wiper fluid!

Most modern cars will tell you when your wiper fluid is running low.

You can buy wiper fluid or a dilution (which you add to water) at your local gas station.

Don't be tempted to just put water in! It is important to include a wiper fluid additive, as the additive contains antifreeze, which will stop it from freezing in the winter months. It also has cleaning agents that will help clean your windshield.

TIRE PRESSURE

Why Is Tire Pressure Important?

Tire pressure is the force exerted by air against the inner walls of a tire. Driving with a tire that is substantially under- or overinflated can result in tire failure. This can be dangerous if you're driving fast, as you can lose control of the car. The wrong tire pressure can also mean that the car's braking capability is dramatically reduced, which, if you are driving at high speeds, can also make you lose control.

Having an underinflated tire shortens the life of the tire and means your car has to work harder. This ultimately uses more fuel unnecessarily— more money = less shopping!

How Do I Know What My Tire Pressure Should Be?

You can find out what tire pressure your car needs by consulting your owner's manual. However, sometimes it is written in the glove compart- ment, on the driver's doorjamb, or inside the gas cap. Some car makes require different air pressure in the front tires compared with the back.

Most gas stations have a tire pressure gauge in the service area that you can use.

Note: Tire pressure needs to be checked when tires are cold, so ideally check them after a very short journey (e.g., from your house to the nearest gas station).

Checking Your Tire Pressure

* Remove the cap. These can get lost, but the pressure should be okay without them.

* Put the top of the pressure gauge around the seal. (1)

* Make sure the gauge completely covers all parts of the seal and no air is escaping (i.e., there is no hissing sound). (2)

* Read the pressure levels. (3)

* If the reading on the screen is the same as the pressure your tires should have, then no further action is required on this tire. Continue to the next tire and repeat.

* If the pressure reading you are getting is less than the level it is supposed to be, you need to add more air. To add more air to your tires, increase the pressure gauge to the amount required for your tire. (4)

* Return to your tire and ensure that the gauge completely covers the valve; the system should add the right amount of air to your tire. (5)

* Once you have finished, replace the cap on the valve (if you can find it!).

① PUT PRESSURE GAUGE AROUND SEAL

② ENSURE GAUGE COMPLETELY COVERS ALL PARTS OF THE SEAL

③ READ PRESSURE LEVELS

④ INCREASE PRESSURE GAUGE TO AMOUNT REQUIRED

⑤ RETURN TO YOUR TIRE AND ALLOW GAUGE TO ADD AIR

TIRE CONDITION

You should inspect the condition of your tires every two weeks, while you are checking the air pressure. Look for any excessive wear, cracks, bulging, or deep cuts. It is actually illegal to drive with badly damaged tires.

TREAD DEPTH INDICATOR

You also need to check the tread on the tire. The tread is the grooved surface on your tire that helps the tires grip the ground. If the tread wears down or becomes bald, this can be very dangerous, as the tires won't be able to grip the road properly and you could lose control of your car.

Most countries have laws specifying a minimum tread depth that your tires must have. It is usually around 1.6 mm.

Most new tires have a built-in tread depth indicator. When the tires are worn down to this level, they need to be replaced.

TO GET THE BEST VIEW FOR CHECKING THE TREAD DEPTH INDICATOR, TURN THE WHEELS OUT.

EVEN WEAR

UNDER-INFLATED

OVER-INFLATED

TOE IN

What Your Tires Are Trying to Tell You

The way tires wear can usually tell you a great deal about your car's condition. Healthy tires wear evenly. The tread in the middle should be slightly thicker than the edges.

Underinflated

If the outer edges of the tire are more worn down than the inner edges, this means your tires tend to be underinflated.

Overinflated

If the tread in the center of the tire is more worn than the outer edges, you usually drive with your tires overinflated.

Toe In

If your tires are worn more on one side than on the other, then you have what is known as a toe in, and you need to get your wheels realigned.

Tire Rotation

One way of preserving the life of your tires is to rotate them every five thousand to seven thousand miles. Ask your mechanic to do this for you while you are getting your car serviced.

WINDSHIELD WIPERS

As you use wipers, they begin to wear down. They should be changed once or twice a year, depending on their usage. Some manufacturers recommend every six thousand to ten thousand miles.

You can tell when your wipers are wearing down because they begin to smear the glass or make a noise when they're in use.

You can purchase windshield wipers from most gas stations, or you can go to an auto shop. You can also go to your car's dealer center and purchase the manufacturer's recommended wipers—although those will probably be more expensive.

YOU CAN REPLACE THE WHOLE WIPER BLADE OR JUST REPLACE THE RUBBER INSERT, WHICH IS SLIGHTLY TRICKIER.

① DEPENDING ON YOUR WIPER BLADE, YOU WILL HAVE TO EITHER PULL OR PUSH YOUR WIPER FROM THE MAIN BODY TO REMOVE IT.

② INSERT THE NEW WIPER. YOU WILL GENERALLY HEAR A CLICK AS IT GOES INTO PLACE.

③ REPLACE THE WIPER ON THE WINDSHIELD.

WHEEL ALIGNMENT

Wheel alignment and tire balancing are two totally different things, but people often get them confused.

Wheel alignment means adjusting the angles of the wheels so they are perpendicular to the ground and parallel to each other. Aligning your wheels maximizes the life of the tires and maximizes gas mileage.

IF YOU ARE EXPERIENCING ANY OF THESE PROBLEMS, YOU NEED TO TAKE YOUR CAR TO A MECHANIC OR A TIRE CENTER TO GET YOUR WHEELS ALIGNED.

Signs Your Wheels Are out of Alignment

1. Uneven or rapid tire wear.
2. Pulling or drifting when you are driving in a straight line.
3. The spokes of your steering wheel are off to one side when you are driving on a straight road.

TIRE BALANCING

If you experience vibrations from your steering wheel, seat, or floorboard when you are driving at high speeds, it is likely that you need to get your tires balanced. A tire can become unbalanced just like a washing machine during a spin cycle. This is another operation that needs to be carried out by professionals. You can take your car to your friendly local mechanic or tire center.

BODYWORK

Scratches and scrapes are all part of the driving process. Some people scratch and scrape more than others, but if you leave a scratch untreated, rust can set in. This destroys the metal underneath, and you will have to get the whole panel replaced, which is pricey!

Replacing the entire section will probably leave you out of pocket even more when you try to resell the car. Scrapes and scratches reduce the resale price of the car. Respraying parts can be expensive, but it's worth it.

When you are having your car sprayed, don't be afraid to call several places to get a quote. You may find the quotes vary greatly, and you might get yourself a great deal!

SPOTTING PROBLEMS

A Few Things to Look Out For

Watch out for leaks or stains under the car. If you notice your car is leaving any patches of fluid on the ground, it is likely you have a leak and need to take your car to a mechanic.

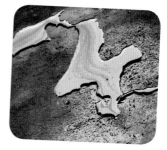

If you are constantly having to replace fluid (e.g., oil), this means you could have a leak.

Be aware of all of your controls when you are driving, and if anything feels a little off, make a note of it and inform your mechanic. For example: stiff gear changes, an unusual sensation in your pedal, your steering feels different, you hear strange noises.

WHAT IS A CAR SERVICE?

Your car is a sophisticated piece of machinery that relies on everything working as it should in order to obtain peak performance and safety. When car manufacturers are designing and testing a car, they come up with recommended procedures that need to be carried out in order to ensure the car they have produced works properly. Filters, fluids, belts, brake pads, and spark plugs, to name a few, all need to be changed after a certain amount of time or mileage. Your car comes with a service manual that will tell the mechanic what procedures he needs to carry out depending on what mileage your car has reached.

Check your owner's manual to find out how often you need to bring your car in for service. Don't try to avoid it! Continually paying small amounts for ongoing service will reduce your chances of having to pay out whopping sums that you can't afford when major things go wrong.

Visiting Your Friendly Local Mechanic

It's safe to say that a trip to the mechanic is not a favorite passtime for many women. You nervously hand over your car, trying to recall the last time you had it serviced. You return a few days later only to be told that something is wrong with it, most of which you don't understand. Then you hand over your money and say goodbye to that two weeks in Spain with the girls. Then there's that nagging suspicion that—could it be?— Mr. Local Mechanic is taking advantage of the fact that you're *female*?!

Here Are a Few Tips for Trying to Minimize the Emotional Trauma:

1 Go in sounding knowledgeable. Use this book to help you determine what's wrong with your car so you can tell the mechanic what the problem is, and not the other way around. If possible, use some mechanical terminology to convince him.

2 Get recommendations. Ask friends to recommend a reliable and trustworthy mechanic.

3 Make sure the mechanic is qualified—don't be tempted to give your car to a friend's brother who'll do it for less.

4 If you need a lot of work done, don't be afraid to haggle or call several other garages to get a comparison quote.

5 Agree on the work you want the mechanic to do, and on the price. Sometimes when a mechanic starts working on a car, he finds other things wrong with it. Tell the mechanic to call you before he carries out any additional work that will increase your bill when you collect your car.

6 The purpose of this book is to empower girls when it comes to their car. However, as a last resort, if you don't think you will be able to talk the talk in this situation, then it might be a good idea to bring a trusty I-know-all-about-cars friend with you to do the talking on your behalf. It's good to make them feel useful!

WHO DO I GO TO?

Repairing cars has become a very specialized science, so when something goes wrong with your car, make sure you contact the right person.

- ✳ CRASH REPAIR CENTERS (aka panel beaters): repair any damage done to the exterior of your car.
- ✳ MECHANICS: carry out all work involving your engine.
- ✳ TIRE CENTERS: sell tires and perform wheel alignment and balancing.
- ✳ AUTO ELECTRICIANS: fix any electrical problems (e.g., alarms), problems with your battery, or problems with your lights.
- ✳ CAR DEALERSHIPS: even if you didn't buy your car from an authorized dealership, occasionally when something goes wrong, you will have to deal with a certified dealership for your make of car. You can find your nearest dealership by looking up your car make in the phone book or online. You will need to go to a dealership if you lose your car keys, have problems with your alarm, or need to buy a specialized part for your car. Car dealerships can also carry out servicing, but they tend to be expensive compared with other mechanics.

One of the best ways to get good auto repair is to look for a technician or other service professional with ASE certification. The easiest way to find one is to look for the ASE sign posted at the facility or in its Yellow Pages, newspaper, or online ads. Most of these establishments proudly display the ASE certificates earned by their technicians in their office or waiting room. Finally, ASE-certified technicians often wear the ASE Blue Seal patch on the shoulder of their uniforms and are more than willing to show you their credentials.

Chapter Four

CAR DOCTOR

CAR DOCTOR...
QUESTIONS AND ANSWERS

Starting Problems

Q When I try to start my car, the engine won't start but there is a clicking noise. What should I do?

A It sounds like your battery is dead. Turn to page 98 to find out how to recharge it. If your battery is more than three years old, it may be at the end of its life and may need to be replaced.

Q I have tried to start my car, but it won't start. The headlights are working full beam, though. What do you recommend?

A It could be a problem with your ignition and not your battery. In this event, jump-starting may not help. You will need to call a mechanic.

Pedal Problems

Q When I put my foot on the brake, I feel a pulsating/vibrating sensation. Can you explain?

A With ABS, a pulsating effect is normal when the driver is braking hard. If you don't have an ABS system, then it could mean that the brake discs are damaged or warped. Your local mechanic should be able to replace or refinish a damaged disc for you.

Q Sometimes when I am accelerating, there is a squealing noise. What do you think that could be?

A It could indicate that you have a loose alternator drive belt. This can lead to other problems, such as engine overheating or a dead battery, so take it to your mechanic and have him tighten the belt for you.

Q Lately when I have been pressing my foot against the brakes, it feels more springy than solid. Is this something I should be worried about?

A As brakes are a vital part of your safety system, they should not be ignored. It is more than likely that air has gotten into your brake system. Not only do your brakes need to be bled, but your mechanic also needs to determine how the air got into the system in the first place.

Q When I try to accelerate, the revs on my rev counter increase but my car doesn't actually speed up. What is happening?

A Your clutch has probably slipped, so it needs to be overhauled by a mechanic right away.

Q Lately, when I put my foot on the accelerator pedal, it is quite springy, and when I try to put my car in first gear, I hear a grating noise. Do you know what's wrong?

A I would say that air has gotten into your transmission system and it needs to be bled. Be wary of someone who tells you need an entire (and expensive) overhaul, as this should not be the case.

Exhaust Problems

Q My exhaust seems to be making a lot of noise recently. What is wrong?

A There could be a hole somewhere in your exhaust system, and gases are blowing out under the car, rather than out the back of it. This can be a dangerous condition, so take your car to a mechanic or exhaust specialist.

Q I noticed droplets of water are coming from my exhaust pipe. What do you think is wrong?

A Some condensation coming from your exhaust pipe is normal first thing in the morning, as condensation has built up in it overnight. However, if your engine is fully warmed up and you can still see signs of excessive water coming from your exhaust, it could mean there's a more serious problem with your head gasket that will need the urgent attention of a mechanic.

Q Sometimes when I have been sitting in traffic for a while, a cloud of blue smoke comes from my exhaust. What could that be?

A It could be that your pistons, piston rings, or valves are worn and need replacing. It is best to take the car to your mechanic to get it checked out.

Engine Problems

Q I can hear a heavy knocking noise coming from my engine when I am driving. Any suggestions?

A If the noise is coming from your engine, it could be that the bearings on your engine are worn and need to be replaced. Unfortunately, this is an expensive but essential procedure.

Q I can hear a hissing noise coming from my engine when I am driving. What is that?

A It sounds like you have a leak in either the intake manifold or one of its associated parts. In this case, you will need to get your gasket tightened or replaced by a mechanic.

Q I am driving my car normally, but it feels like there is a lot less power. Can you help?

A You probably have faulty ignition timing. You need to take the car to a mechanic to get it checked out.

Q My engine vibrates violently and seriously lacks power when it's running. What do you think could be wrong?

A It sounds like a misfire in one cylinder—possibly, one ignition coil needs replacing. You will have to take it to a mechanic, but luckily, fixing it is not a major undertaking.

Miscellaneous Problems

Q I have a small crack in my windshield. Will I need to get the whole window replaced?

A Maybe not. There are windshield specialists who deal with these issues. You can find them in your local phone book or online. If the hole is smaller than a quarter, is two inches in from the edge, and is not in the driver's immediate vision, then they should be able to repair it for you without having to replace the whole windshield. The sooner you get the crack treated, the better, as it can quickly increase in size.

Q I have just put gasoline into my diesel engine. What should I do?

A If you have put less than a quarter of a tank of gasoline into a diesel engine, then you should be okay. Ensure you fill the rest of the tank fully with diesel. When you drive, the engine might smoke a bit, but it should be fine. If there is more than a quarter of a tank of gasoline in your engine, then do not turn the engine on. Get it towed to a mechanic and have him drain the fuel tank for you.

Q Help! I have just put diesel into my gasoline engine.

A Putting diesel into a gasoline engine is much more serious than the other way around. Do not attempt to turn on the engine. The car needs to be towed to a mechanic who can drain the fuel tank and flush the system.

Q The lights in my car are constantly blowing and needing to be replaced. Why is that?

A It could be that your alternator voltage is set too high. The situation can be quite dangerous, as the battery gives off explosive hydrogen gas if it is being overcharged, so take your car to your mechanic so he can attend to it.

Q There is a knocking noise coming from my wheels when I turn corners. What is this?

A It sounds like the CV joints of your wheel need replacing. Your local mechanic can easily perform this repair.

Q Recently, when I have been driving over speed bumps, my car bounces a lot more than usual. What should I do?

A I would say that you need to get new shock absorbers. A mechanic can install them; some tire centers now provide this service, too.

Q While I was doing my maintenance checks, I noticed an excess of white powder around my battery. Is this normal?

A This may indicate that your alternator is overcharging the battery, which can be quite dangerous. Take your car to your local mechanic so he can check the charging rate of your alternator and rectify or replace it as needed.

Chapter Five

CLEANING YOUR CAR

HAVE YOU EVER SEEN A FORTY-YEAR-OLD CLASSIC CAR THAT looked like it was just out of the showroom or a ten-year-old car that looked faded and drab? What caused the difference? You can bet the person with the classic car was out every month cleaning and putting protective products on their car to prevent the aging process.

Driving in a nice, clean car not only makes you look and feel good, but also helps increase the resale value of the car when it is time for you to sell. Get your sponges at the ready and find out how to give your car the proper love and attention it needs to maintain its showroom look.

CLEAN AND PROTECT

Did you know that not only are you supposed to clean the surfaces of your car, but you are also meant to apply a protective coating to them?

The paint on your car is exposed to all sorts of elements—acid rain, pollution, acidic bird droppings, sap from trees—that corrode the paint, making it lose its color and shine. Similarly, the vinyl in the interior of your car is constantly exposed to the sun's harmful UV rays, and unless it gets a protection treatment every now and then, the vinyl will begin to fade and look aged.

Ideally, your car should get a full valet (deep clean) twice a year and then a quick clean every month—vacuum, exterior wash, wax.

Getting Someone Else to Do the Dirty Work

There are professional car valet companies that will do the work for you, but it's not cheap. If you choose to go the valet route, before you hand over your car, don't be afraid to ask them what's involved, and then when you come to collect it, make sure they have done the work they said they would do. If the price included a wax, feel the surface of your paint to make sure it was done.

The brushes used in commercial car washes can be tough on your paint job, and that is why some people prefer to go to "brushless" or "hand wash" car washes, which are softer on your car.

First things first. The cleaning agents in dishwashing liquid are designed to remove animal fats from ceramic plates. Car shampoos are designed to remove all the harmful substances your car's paint is likely to come into contact with, and many of them come with gloss enhancers. Which one are you going to choose to clean your car with? If you have decided to take on the task of cleaning the car yourself, take a trip to your nearest car-care shop and invest in good-quality specialty car-cleaning products. Once you are stocked up, the products should last you several cleans.

Car Cleaning—Shopping List

- Plastic bucket
- Car shampoo
- Carpet cleaner
- Baking soda
- Special citrus degreaser for insects and tar that just won't shift
- Proper wash mitt, pad, or sponge
- Lots and lots of clean 100 percent cotton strips of cloth (tear up those old T-shirts!)
- Chamois
- Car wax
- Vacuum cleaner
- Upholstery cleaner
- Vinyl cleaner and protectant
- Small paintbrush (for getting into tricky corners)

AND IF YOU HAVE LEATHER SEATS:

- Leather cleaner
- Leather conditioner

Car Wash

⊛ Read the instructions on your car shampoo to make sure you mix the right ratio of shampoo to water. When it comes to car shampoos, less is more, so don't overdo it.

⊛ Clean the car in sections (i.e., hood, trunk, and sides).

⊛ Start cleaning from the top down.

⊛ If at any time grit gets into your wash mitt or sponge, remove it immediately as it can scratch the paint.

⊛ Be careful of bird droppings. They often contain seeds that can scratch the paint. If you come across a bird dropping, gently blot it clean with a mix of one tablespoon of baking soda and a mug of warm water.

⊛ Try to dry the car as soon as possible after you have washed it.

⊛ After you have rinsed the car, to dry, gently blot the water away from the surface.

⊛ Use normal window cleaner to clean the windows.

Wax

When the car is fully dry, it's time to wax!

Applying a wax to the surface of the car protects it from all the elements that try to attack it on a daily basis: acid rain, UV rays, pollution, insects, road tar.

Wax should be applied only to a clean surface.

- ⊛ Wax should be applied sparingly, so use your fingers or an applicator pad to apply a small amount.

- ⊛ Rub the wax into the car in a linear motion, preferably in the direction air moves along your car when you are driving.

- ⊛ Work the wax into the surface until all you can see is a light haze.

- ⊛ When all the wax has been applied, get a clean, 100 percent cotton cloth and buff (polish).

- ⊛ When you have finished the first buff, go back around the car again, buffing the areas where you can still see wax.

- ⊛ Stand back and be proud of your lovely, shiny car.

CLEANING THE INTERIOR

- ✳ Take everything out of your car.
- ✳ Remove floor mats.
- ✳ Vacuum as many surfaces as you can.

Upholstery

- ✳ Wet a clean cloth, wring it out, and apply some upholstery cleaner.
- ✳ Clean the upholstery with the cloth.

BE SURE NOT TO SOAK ANY OF THE UPHOLSTERY, OR IT WON'T HAVE A CHANCE TO DRY PROPERLY AND WILL LEAVE A MOLDY SMELL.

Carpets

- Bang floor mats against a surface to remove as much dirt as possible.

- Vacuum each mat on the ground.

- Spray mats evenly with carpet cleaner.

- Leave for a few minutes.

- Wipe thoroughly with a dry cloth.

- Repeat the same cleaning process for interior carpets, ensuring that you don't ever let the carpets get too damp.

Vinyl

Your dashboard, inside doors, steering wheel, and sometimes seats and convertible tops are made from vinyl. Vinyl is sensitive to the sun's rays and can fade or become brittle over time. It is essential that you use a cleaning product that has been specially designed to clean vinyl.

- Using a dry cloth, clear away the excess dirt on all your vinyl surfaces.

- Use a small paintbrush to clean the vents.

- Apply vinyl cleaner to a clean cloth and work it into the surfaces.

- Go over these areas again with another clean, dry cloth.

- Apply vinyl protector.

When you have finished cleaning your car, if you can, leave the doors of the car open for thirty minutes to dry it out.

Leather

As leather is a natural product, it isn't as enduring as man-made products, so it needs a lot more love and care than upholstery seats do to stay looking fresh. If you have splurged on a car with leather seats, doesn't it make sense to look after them properly?

* Remove any excess dirt using a vacuum cleaner.
* Use a brush to remove any dirt that may have become lodged in the grooves of the seats.
* Apply special leather cleaner and gently work into the leather.
* Clean away any excess with a clean cloth.
* Apply a leather conditioner to the seats.

Chapter Six

WHEN THINGS GO WRONG

THROUGHOUT YOUR LIFETIME OF OWNING A CAR, AT SOME POINT you will run into car trouble. Here is a checklist of things every girl should keep in her car in the event that she breaks down or gets into an accident.

- ⊛ Tire gauge
- ⊛ Flashlight
- ⊛ Flares
- ⊛ Jumper cables
- ⊛ Jack
- ⊛ Wheel brace
- ⊛ Cell phone charger that plugs into lighter or charger plug
- ⊛ Bottle of water—for engine overheating

WHAT TO DO IF YOU BREAK DOWN
AND YOU HAVE INSURANCE

If your car breaks down and you feel it isn't a problem you can fix yourself, you need to pull over as soon as it is possible to do so.

- ✳ Try to park your car in the safest place possible.

- ✳ If you are on a freeway, pull it as far onto the hard shoulder as possible.

- ✳ Put on your hazard lights.

- ✳ Call your roadside-assistance company.

- ✳ If you are worried that your car could be hit by oncoming traffic, when it is safe to do so, get out of the car, along with any passengers, and wait in a safe place.

- ✳ If you are able to wait in the car, lock all the doors and roll up the windows.

- ✳ If a stranger tries to talk to you, inform them that you have called for roadside assistance and that you are okay.

WHAT TO DO IF YOU BREAK DOWN
AND YOU DON'T HAVE INSURANCE

If you break down and you don't have any breakdown coverage, the cheapest option is to call a friend with a tow bar and a tow line, which you can purchase from most gas stations, and ask them to tow you to a nearby garage, ideally your local one. If it is after hours, park your car somewhere safe near the garage and ask your friend for a lift home. Then buy them a lovely gift for being so kind.

If you don't have a friend with a tow bar, then you will have to call information and get the number of the roadside-assistance company that is closest to where you have broken down. Be warned, though—these companies can be very expensive. If you don't have the money straight away to pay them, they will take your car to an impound lot and charge you a fee for every day you leave it there. Then you will need to get them to take the car from the lot to a garage, which will cost an additional fee.

FLAT TIRE

So, you're driving along, and your car begins to feel a bit bumpy, and the steering starts to pull to one side—chances are, you have a flat tire. Decide on a suitable expletive, swear loudly, and pull to the side of the road as soon as it's safe!

Changing a flat tire seems to be the thing about cars that girls are scared of the most, but it's really quite simple. You should have a jack in your car that does all the heavy lifting for you. The only heavy lifting you have to do is lifting the tire itself. Believe it or not, it should take you less than fifteen minutes—which is quicker than waiting for someone to come and save you!

What You Need:

① A safety triangle

② A jack

③ A wheel brace

 WARNING—YOUR HANDS ARE GOING TO GET DIRTY!

1. Safety always comes first, so make sure your hand brake is on. If you have a safety triangle, place it behind the car to notify other drivers.

2. If your car has a chock—a piece of metal that you can place under the front wheel to stop your car from rolling—use it.

3. Prepare yourself. Set out your owner's manual, spare tire, jack, and wheel brace within the vicinity of the flat tire. If you have a wheel cover, you must lever this away from the tire first.

4. Loosen the bolts on the tire by turning them about half a revolution in a counterclockwise direction with the wheel brace. Don't be afraid to use your legs if the bolts are very stiff! But don't remove the bolts fully yet.

USUALLY, YOU NEED TO TURN THE NUTS COUNTERCLOCKWISE TO LOOSEN AND CLOCKWISE TO TIGHTEN. A USEFUL RHYME IS "RIGHTY MEANS TIGHTY, LEFTY MEANS LOOSEY"—IT HELPS PEOPLE REMEMBER WHICH WAY TO TURN THE NUTS.

S Next, you need to locate the "jacking point" on the car. This is a reinforced part on the underside of your car that you will place the jack under. This will take the weight of the car when the jack is raising it up, without damaging the undercarriage of your car.

You can look or feel to find the jacking point. Alternatively, consult your owner's manual, which should tell you where it is located. You need to use the jacking point that is closest to the tire you are changing.

6 Place the jack on the ground underneath the jacking point.

IF YOU HAVE ALLOYS, YOU MAY NEED TO REMOVE A SPECIAL BOLT.

CONSULT YOUR OWNER'S MANUAL AND FOLLOW THE INSTRUCTIONS FOR REMOVING ALLOYS.

7 Turn the jack in a clockwise direction.

8 Keep turning the jack until the tire you are changing is lifted a few inches off the ground.

9 Using your fingers, the wheel brace, or the nut key, unscrew the nuts fully.

10 When all the nuts have been removed, lift the wheel away from the car.

11 Replace the old tire with the spare. Warning—it can be quite tricky to align the holes in the tires with the studs!

⑫ Fasten the nuts in place as tightly as you can with your fingers.

⑬ Lower the jack. When the car has been lowered to its normal level, fully tighten the nuts with your foot until the wheel brace cannot turn any more. It is better to tighten the nuts in a star-shaped pattern than in a circle. This means tightening the nut opposite the one you have just tightened.

NOTE: IF YOU HAVE A SPACE-SAVER TIRE, IT IS MEANT ONLY FOR TEMPORARY USE—ENOUGH TO TAKE YOU TO A GARAGE OR TIRE CENTER. DO NOT TREAT THIS AS A NORMAL TIRE.

DEAD BATTERY

ALWAYS TAKE CARE WHEN HANDLING BATTERIES, AS THEY CONTAIN TOXIC CHEMICALS AND INCORRECT HANDLING CAN CAUSE SERIOUS INJURIES. KEEP CHILDREN AND NAKED FLAMES AWAY FROM THE BATTERY.

If you turn your ignition key and all you hear is a constant clicking or tapping noise, it means you are having problems with your battery.

The first thing to do is check that your battery isn't covered in dirt. If it is dirty, clean the points on the battery, as doing so might solve the problem.

If you do this and it still doesn't work, then your battery is dead and you will have to jump-start the car. But not to worry—it's really quick and easy once you know how.

① You will need to find a kind person who has a car with a fully charged battery and a good pair of jumper cables.

② Position the working car so it is front to front with the dead car. If it is not possible to line up the cars in this way, line them up so the batteries are close together and the jumper cables can reach from one battery to the other.

3 Make sure both cars are positioned in park and both engines are turned off.

4 Make sure the radio and lights on the dead car are switched off.

5 Your battery may have a cover over it. If it does, remove it.

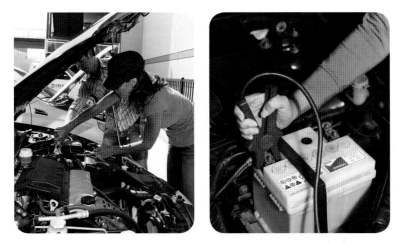

6   Attach the red cable of the jumper cables to the positive point on the dead car's battery. Usually, the positive point is often red in color and has a plus sign on it.

7   Attach the other end of the red cable to the positive point of the working car.

8   Attach the black cable of the jumper cables to the negative point on the dead car's battery. The negative point is usually black and has a minus sign on it.

9   Attach the other end of the black cable to the negative point of the working car's battery.

1 0 Let the working car start up and rev its engine (in neutral) for about twenty seconds.

1 1 After twenty seconds, turn on the ignition of the dead car, and it should start for you.

1 2 When the dead car's engine begins to work, rev its engine (in neutral) and keep it running for a while to recharge the battery.

BATTERIES HAVE A LIMITED LIFE SPAN—ON AVERAGE, 3.5 YEARS. AFTER 3.5 YEARS, ONCE A BATTERY STARTS TO FAIL, IT VERY RARELY RESPONDS TO RECHARGING AND YOU WILL NEED TO REPLACE IT.

03 Now, carefully remove the jumper cables from both engines.

04 It is best if you drive the formerly dead car around or leave the engine running in neutral for a while so it can recharge itself.

NOTE: JUST BECAUSE YOUR BATTERY IS DEAD DOESN'T NECESSARILY MEAN IT NEEDS TO BE REPLACED. IT MAY JUST MEAN THAT YOU HAVE BEEN DOING LOTS OF SHORT TRIPS OR HAVE LEFT IT DORMANT AND IT HASN'T HAD A CHANCE TO RECHARGE ITSELF.

If you're still having constant battery problems but you are sure that there is nothing wrong with your battery, then you could have an alternator problem. Dim headlights can be a sign of this issue. Seek advice from your mechanic.

ENGINE OVERHEATING

If you are driving along and steam starts coming out of your engine and your temperature gauge reads anywhere on the HOT spectrum, pull over as soon as it is possible and safe to do so. Do not wait until you reach a gas station or garage—you could cause a huge amount of damage to your engine.

(1) Put on your hazard lights to notify other drivers you are in trouble.

(2) Open the hood (look under "Hood" in your owner's manual if you don't know how to do this).

(3) Do not touch any part of the engine, as it will be extremely hot and could burn you.

(4) Be patient—your engine can take a long while to cool down to a level at which it is safe to touch.

(5) Turn on the heat in the car full blast. This takes heat away from the engine.

6 You will need to add coolant, but you can use water in an emergency situation. Either of these will need to be lukewarm, as putting cold water in a hot engine can cause it to crack. You may need to find the coolant or water while you are waiting for the engine to cool down.

7 If you hover your hand above the radiator, you should be able to tell when it is cool enough to touch. Even then, you might want to cover your hand with a rag or material while you are opening it. It is likely that steam will gush out of the radiator, and you do not want to get scalded.

8 At arm's length, slowly add the lukewarm coolant or water to the radiator cap (see owner's manual for location if you are unsure).

9 Drive the car only if the temperature gauge returns to normal.

10 You need to take your car to a mechanic ASAP, so do not plan to drive long distances.

11 If your car begins to overheat again, pull over and call your local recovery company. Your car is likely to have serious problems, and driving it any further can cause terminal damage.

LOST KEYS

If you lose your keys and your car was made before the year 2000, you might be in luck. If you can find out the vehicle identification number of your car (consult your owner's manual to find it), then call a locksmith or your local dealer, and they should be able to cut a new key for you.

However, if your car was made after the year 2000, it is more than likely that losing your key is going to turn into an expensive and troublesome experience. Most car keys made after 2000 have a special microchip, called a transponder, in the body of the key.

When you put your key in the ignition, there is a device that reads the microchip and will allow the key to turn only if it has this special microchip in it. So if you lose your keys, you are going to have to get a new key with a microchip. Locksmiths do not have the capability to program these microchips, so you must get the new key programmed at a car dealership.

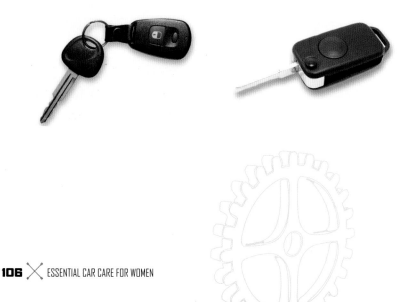

What to Do Next:

1 Unless your local dealership has a service that can come to you, you will have to get your car towed to your nearest car dealership—expense number one.

2 They will cut the key for you—expense number two.

3 Then they will program the key for you—expense number three.

4 Note that car dealerships are open only during normal working hours, so if you lose your keys outside of these hours, there is nothing you can do until the dealership reopens.

NOW THAT YOU CAN APPRECIATE HOW TROUBLESOME AND EXPENSIVE LOSING YOUR KEY IS, TRY TO MAKE SURE YOU KEEP YOUR SPARE SET OF KEYS SOMEWHERE SAFE. REPLACING A LOST KEY NOWADAYS WILL DIG INTO YOUR VACATION FUND!

FAULTY ALARM

Occasionally, your car alarm will go off for no reason. This suggests that you have a fault in your alarm system. The siren will sound for four or five hours before the battery dies. By then, you'll have a grumpy neighbor *and* a dead car battery to add to your woes!

It is unlikely a mechanic will be able to solve the problem. If your alarm is faulty, you will have to get it repaired by an auto electrician or your car dealership.

WHAT TO DO AFTER A CAR ACCIDENT

Being involved in a car accident, no matter how small, can be quite an upsetting experience, so the most important thing is to stay calm. Even if there is only a small amount of damage done to either of the cars, the repair costs could be surprisingly expensive. The person at fault is the person responsible for paying any damages, but won't have to pay for them personally. This is why we have auto insurance.

In order for both parties' insurance companies to be able to handle the situation correctly, you need to carry out certain steps. If you don't, then you risk ending up having to pay for the repairs yourself—even if the accident wasn't your fault.

If you have been involved in a car crash, *do not* admit any liability (that it was your fault), even if you were in the wrong. When you are dealing with the other driver(s), handle the situation as factually as possible, and try to make an accurate record of what happened. You can discuss a situation factually without apportioning any blame to anyone. Leave it up to your insurance company to decide who's at fault.

Accident Protocol

1 Stay calm. Take lots of deep breaths and try to relax your mind.

2 When it is safe to do so, get out of the car and assess the situation.

3 Determine whether or not you need to call the police.

You *don't* need to call the police if:

⊛ the cars have suffered only minor material damage and are still roadworthy. However, you can call the police if you want to.

You *do* need to call the police if:

⊛ anybody complains of physical injury.

⊛ there is a significant amount of damage to either of the cars and they are unroadworthy.

⊛ the crash is the result of an illegal driving maneuver, such as somebody running a red light.

If you call the police, they should oversee the situation for you, but for your own peace of mind, it would be worthwhile to ensure that all of the following points have been carried out.

How to Handle a Minor Car Crash Without the Police

If all parties feel there is no need to call the police, then you will need to do the following:

1. Quickly try to take a few pictures of the crash scene with your camera phone at different angles, before moving the cars. If you don't have a camera on your phone, try to find someone else you can ask. Make sure you get their contact details so you can follow up with them for the photos. If you still can't find a camera, then take the time to sketch out the scene so you have a record of it.

2. When you have taken a record of the scene, move the cars out of the way of the moving traffic.

3. Get a pen and paper. If no one at the scene has them, perhaps you can save the details in your phone. You need to get the following information from the other party and give all the same details to them:

* NAME

* ADDRESS

* TELEPHONE NUMBER

* INSURANCE COMPANY AND POLICY NUMBER (YOU CAN FIND BOTH THESE PIECES OF INFORMATION ON THE OTHER PARTY'S INSURANCE CARD)

* IF THEY ARE NOT THE OWNER OF THE CAR (E.G., A COMPANY VEHICLE OR RENTED CAR), THEN YOU NEED TO GET THE NAME AND CONTACT DETAILS OF THE ORGANIZATION THAT OWNS THE CAR

* THE CAR'S REGISTRATION NUMBER

* DESCRIPTION OF THE CAR (COLOR, MAKE, AND MODEL)

* DATE, TIME, AND LOCATION OF THE ACCIDENT

* NAMES AND CONTACT DETAILS OF ANY WITNESSES WHO SAW THE ACCIDENT

⑷ At your earliest convenience (preferably within twenty-four hours),
go to a police station and report the accident. You need to do this for
insurance purposes.

⑸ Contact your insurance company and tell them what happened.

Chapter Seven

DRIVING IN THE SNOW

IF YOU LIVE IN AN AREA THAT IS SUBJECT TO SNOW, OR YOU ARE planning a ski trip to the mountains, don't get in to the car without being prepared for extreme winter conditions. For the uninitiated, it is a whole new driving experience.

When snow grips the ground, or black ice forms, it can be like trying to drive your car on an ice-skating rink: You run an extremely high risk of skidding uncontrollably, and this can cause you to crash.

Thankfully, there are solutions to help you navigate your car safely through snow and ice. If your tires have added grip, it will greatly reduce your risk of skidding. However, while it does reduce the risk, it doesn't eliminate it. Whether you winterize your car by using snow tires, or add chains to your standard road tires when you know you'll be exposed to snowy conditions, you must always drive with extreme care through snow or ice conditions.

This section outlines the options available to you to prepare your car for snow.

SNOW TIRES

These can also be referred to as "traction tires." They have a deeper tread compared to traditional tires, providing extra grip and traction on snow and ice. Additionally, they are manufactured using a special compound that prevents them from getting hard in the cold conditions. However, it is not advised to keep these tires on all year round, as they become too soft during the warmer weather and wear out quicker.

Advantages

If you live in an area where there are long periods of snow, having snow tires saves you the hassle of having to put on and take off chains.

Disadvantages

This means having to buy a whole new set of tires, and they don't come cheap. They range in cost from $300 to $1500, and you may also need to pay to have them installed. Furthermore, when you are driving with these tires on dry roads, they typically wear out faster than traditional tires.

SNOW CHAINS

These chains are wrapped around the tires and provide the extra traction needed to help standard tires grip the road better.

Advantages

Snow chains are a cheaper option compared to snow tires, and sometimes they provide a better grip than snow tires.

Disadvantages

When a blizzard hits and it is freezing cold and snowing outside (or worse, the blizzard occurs while you're driving at night), the last thing you want to have to do is pull over to the side of the road and try to figure out how to put on snow chains.

Another disadvantage to using them is that as soon as you reach dry, snow-free roads, you'll need to take them off immediately to prevent them from breaking and damaging your car or tires.

Despite these disadvantages, they are still the most popular option if you don't spend a great deal of time in areas that experience lots of snow.

Learning how to install snow chains correctly is key to ensuring safe travels, should you find yourself driving in snow or icy conditions. The DMV recommends doing a test run of installing and removing snow chains before hitting the snow. Stranded on the side of the road during a blizzard is not the best time to learn a new skill.

DIFFERENT TYPES OF SNOW CHAINS

When it comes to buying snow chains, there are a few different options.

Traditional Snow Chains

Traditional snow chains are made of heavy metal links that when rolled out create a ladder shape, which you wrap around your tire.

Compared to other types of chains, they can be slightly more bulky and cumbersome to put on.

Diamond Chains

Diamond chains are a variation of the traditional metal-link snow chains, but instead of being shaped like a ladder, they are crisscrossed to form a diamond shape. These are considered a better option to the ladder pattern because they provide more stability and less vibration and noise when you're driving with them on.

Cable Chains

Cable chains are a more modern version of the traditional chain. The cable is made of heavy-duty aircraft steel wrapped in a steel case. They are also usually lighter than traditional metal-link chains.

BUYING SNOW CHAINS

Chains typically come in sets of two, and they are meant to be put on either the front or rear tires of your car, depending on whether you have a "front wheel drive" automobile or "rear wheel drive." However, some people like to put chains on all four tires for extra stability. Obviously, having to buy and install two sets of chains is more expensive and more work.

You can buy snow chains online, in a car accessory shop, or in department stores such as Walmart, Sears, and Target—particularly in areas that get snow.

Unfortunately, there is not a "one size fits all" snow chain that you can buy for your car. Different size tires require different size chains. Additionally, the make and model of your car may restrict the types of chains or cables you can use. Before you go to the store, check the owner's manual for your car, under the section "snow chains" or "traction," to see if there are any restrictions and/or recommendations on what to use. Also, let the store assistant know if you have modified tires on your car.

If you want to be super cautious, other safety items you can purchase include the following:

- ✳ A headlamp is a very useful device to keep alongside your chains because it allows you to keep both hands free if you have to install chains in the dark.

- ✳ A high-visibility reflective vest can be a lifesaver, since it makes you visible to other drivers on the road, particularly if you are installing chains during a blizzard or in the dark.

Ask your sales associate if the chains come with additional tighteners. The tighter you can pull your chains when you are installing them, the safer and more effective they will be.

PUTTING ON SNOW CHAINS|CABLES

The Right Time to Put Them On

It is not good for your tires, the chains, and the road for you to put on chains when there is no snow on the road. However, it is best to put on your chains before conditions get too bad—typically as soon as the snow starts sticking to the ground or there is ice on the road.

 Park your car in a safe spot at the side of the road with your tires straight on and put on your hazard lights. If you have a headlamp and high-visibility vest, put them on.

2 Lay the snow chain out on the ground beside the wheel you are about to put it on and remove any twists or kinks.

3 Make sure the side chains are as straight as possible and that the cross bars of the chain are straight too.

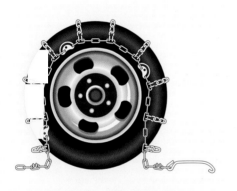

4 When the chains are laid out properly, lift them with care—to avoid tangling them—and place the center of the chain at the top of the tire. Be sure there is an even amount of the chain hanging down either side of the tire.

5 Adjust the chains so that the "ladder" rungs are as straight as possible and they fall evenly on both sides, all around the tire.

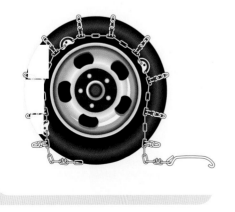

6 Tuck both ends of the chain underneath the tire, but make sure that the fasteners are facing away from the tire (e.g., the outward part of the outside chain is facing the road, and the outside part of the inner chain is facing toward the other side of the car.

7 Repeat this action on the opposite tire.

8 Get into your car and slowly drive about six to nine inches forward.

9 Go back to your tires and on the inside of the tire connect the clasp to the chain. This action is like trying to put on a bracelet really tightly. You need to attach the clasp to the furthest ring you possibly can. If it is easy to clasp the chain, you are probably not pulling hard enough! Sometimes, it can be tricky to tighten the clasps/chains on the inside of the tire. It can help to turn your wheels out via the steering wheel to give you more access.

10 Repeat this action or both the inside and outside of the chains/tires.

11 Repeat the same process on the opposite tire.

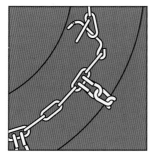

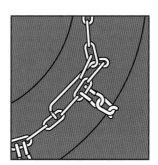

You know your chains are tight enough when you try to pull them away from the tire and there is no "give." If you can still feel some give, then reopen the fasteners and try and get them on to another loop.

Some chains come with one or many chain tighteners (see image). They also come with a wench to tighten them, which you will need to insert into the fastener and twist in order to tighten the chains as much as possible.

Alternately, your chains may also come with a set of "tightening chords," which have hooks on them. With these, you need to put one hook on the outside of one part of the chain, then stretch it across the wheel and place the hook on the opposite side of the chain. You do this on the inside and outside of the tire to ensure the chains are as tight as possible.

Chapter Eight

BUYING A CAR

MANY OF US KNOW WHAT IT FEELS LIKE TO BUY SOMETHING ON impulse. We see something we really like and we want it *now*. But it's important to step back, take a breath, and ask ourselves if we can actually afford it. Buying a car is a big commitment and one that you need to take some time making a decision on. If this is your first time buying a car and you just love the new Mini Cooper, you might be a bit disappointed when you investigate things further and find out that you can't actually afford it. It is not just the cost of buying the car itself that comes into account. There are lots of other, very significant extra costs that come with owning and running a car. Before you even begin to start looking at cars, the first thing you need to do is figure out how much you have to spend.

FIGURING OUT WHAT YOU CAN REALLY AFFORD

The best thing to do is look at your monthly income and see how much you can afford to set aside for the luxury (yes, it is a luxury) of owning a car. Once you have determined how much you have to spend, read through the next few points to become aware of the real cost of owning and running a car so you can choose the car that won't leave you financially crippled.

THE CAR ITSELF

So you see a car you really like. How much does it cost? Usually banks offer car loans for a sixty-month period, so divide the price of the car by sixty. This is roughly how much your monthly repayments will be— although in reality, they will be slightly higher, as you will have interest to pay on top of that, but at least you'll have a ballpark figure you can work from. Can you afford to pay that amount each month? If this is the first car you are going to buy and this figure is stretching you, don't even think about it. This is only one cost of driving. If you've owned a car before, you know there are other expenses that come with owning a car. Have you taken those other costs of driving into account yet?

THE OTHER COSTS OF DRIVING

Gasoline

Determine:

- ⊛ How many miles are you planning on driving each month?
- ⊛ How many miles does your car do per gallon?
- ⊛ How much is the price of gas per gallon?

Use the calculation below to estimate your monthly gasoline cost.

Miles per month / Miles per gallon = Monthly gas cost

Registration

Look up your state's DMV website to find out how much the registration for the car you are interested in will cost each year, and divide that by twelve to get a monthly figure.

Insurance

Call up the insurance companies and get a quote on your insurance for the car. They will often give you an annual and a monthly quote.

Maintenance

Ideally, you should get your car serviced twice a year. Call up your local mechanic and ask him to estimate how much he thinks a service on the car you are interested in might cost.

Unexpected Costs

Unfortunately, when you own a car, you encounter some unexpected expenses. Trips to the mechanic and body shop are never cheap, and you never know when something is going to pop up. Just be aware that you will more than likely have to shell out several hundred dollars every year on unexpected expenses.

Now that you have all your figures at the ready, calculate the total monthly cost of owning that car.

Monthly Cost of Running a Car

1 MONTHLY REPAYMENTS	
2 GAS	
3 INSURANCE	
4 REGISTRATION	
5 MAINTENANCE	
6 UNEXPECTED COSTS	
TOTAL MONTHLY AMOUNT	**$**

IF THE COST IS MORE THAN YOU CAN AFFORD, TRY LOOKING FOR AN OLDER BUT CHEAPER CAR OR A CAR WITH A SMALLER-SIZE ENGINE, WHICH WILL LOWER YOUR GAS, INSURANCE, AND REGISTRATION COSTS.

BUYING NEW, BUYING USED, OR LEASING

Now that you have decided how much you can afford to spend on the car, you need to decide which finance option best suits your needs: buying new, buying used, or leasing.

You are about to part with a significant amount of your money, so you want to make sure you make the right and informed decision and buy a car that suits your personal and financial needs. You also need to ensure that you will get some money back on it when you go to sell it sometime in the future.

But which option is best for you? Outlined here are the pros and cons of each of the different types of strategies.

Buying New

The Advantages

Some people just love the feel of owning a new car: the smell, the pristine condition, having the latest license plates.

One of the biggest advantages of buying new is that you get the manufacturer's warranty. A warranty is a guarantee from the manufacturer that the car will perform the way they say it will. Warranties are usually given for a set amount of years (usually three) or a set amount of miles (say, thirty-six thousand), or whichever comes first.

This means that if you have any component problems with the car during this period, you just take it back to the dealer; they will repair it and you shouldn't have to pay anything.

The Disadvantages

But this all comes with a price. The single biggest disadvantage to buying a new car is depreciation.

Depreciation is the decline in value of your car over the course of its life. Say the life span of your car is fifteen years. Every year, your car becomes worth a little less, until fifteen years later, when it is completely worthless and is fit only for the scrap heap. However, the rate at which your car devalues each year isn't even. It tends to rapidly decrease in value in the first three years and then stabilize after that.

Typically, a car loses 30 percent of its value as soon as it is driven out of the car showroom. So, if you just paid $20,000 for a brand-new car, a week later it is worth only $14,000 (70 percent of the original cost). This can be a great argument for buying a "nearly new" car. Some cars can lose up to 50 percent of their value in the first year!

However, it should be noted that not all cars depreciate at the same rate. Some cars hold their value better than others. So if you have your heart set on buying a new car, then it would be worth your while to investigate how the car you are looking at is likely to depreciate.

Why Do Cars Depreciate at Different Rates?

Think of it as buying a rare, coveted designer handbag that there is a waiting list for. You may have to spend a fortune on it, but you can use it for a year, and then, if you tire of it, you can sell it and you will still have lots of interested parties who are willing to pay you good money for it.

However, if you buy a not-so-expensive handbag by one of the mass-produced brands and after a year you try to sell it on eBay, even if it is in mint condition, you aren't going to get a good price because there are loads of these same bags already for sale!

It's the same with cars. If you buy a mass-produced car from one of the well-known car manufacturers, when the time comes to sell your car, there are more than likely going to be lots of other people selling the same car at the same time as you. Alternatively, if you have a rare or limited-edition car, when it comes to selling it, you could have lots of interested parties all outbidding each other to buy your car, therefore bumping up the price.

How Can I Figure Out How Much My Car Is Going to Depreciate?

The best trick is to go online and see how much one-year-old versions of your car are selling for. If you are looking at buying a new Renault Clio, go online and look at what a one-year-old similar Renault Clio is selling for. Subtract that amount from the price you are about to pay for yours, and it will give you an indication of how much your car will devalue in the first year. If you are planning on holding on to it for three years, check what a three-year-old model is selling for. Remember, sometimes it loses most of its value in the first year and then the drop in price can trail off.

Still got your heart set on buying new? If you have factored in depreciation and are still happy to go ahead, then at least you won't get a shock when you go on to sell the car and find out that you will get a lot less than you originally paid.

TIP: IF YOU ARE SELF-EMPLOYED, YOUR CAR CAN BE TAX DEDUCTIBLE, SO YOU SHOULD INVESTIGATE THIS WHEN DECIDING WHETHER TO BUY OR LEASE A CAR.

Consumer Reports

Every year ConsumerReports.com rates the top cars based on price, safety, efficiency, etc. When you are buying a car, it is a good idea to check these reports to help you make your decision.

In 2011, the top ten cars published by ConsumerReports.com were:

1. Honda Fit
2. Toyota Corolla LE
3. Mini Cooper (base, manual)
4. Volkswagen Golf
5. Toyota Prius IV
6. Hyundai Sonata GLS (4-cyl.)
7. Hyundai Elantra GLS
8. Toyota Camry LE (4-cyl.)
9. Toyota Camry Hybrid
10. Toyota RAV4 (base, 4-cyl.)

LEASING

What Is Leasing?

Leasing is when you rent a car from a car-leasing company for a specified amount of time with an agreed preset amount of miles at a fixed monthly rate.

At the end of your lease contract, you give the car back to the car-leasing company, pay any additional costs you may have incurred (e.g., by going over the agreed miles or by causing any excessive wear and tear), and that's it!

The difference between leasing and buying a car is like the difference between renting and buying an apartment. If you rent an apartment, it is not as big a commitment as buying, but when you have finished with it, you have nothing to show for it.

The Advantages of Leasing Are:

(1) You can drive a newer, more prestigious car for your budget than if you were to buy a car outright.

(2) Your monthly repayments will be lower than if you bought a car.

(3) As only newer models are available to lease, they tend to still be under the manufacturer's warranty, so you should have less expensive maintenance costs.

(4) The up-front costs you have to pay will be lower than the amount you would have to spend if you were buying a car.

(5) You can ask for the option to purchase the car at the end of the lease agreement if you feel you might decide you want to keep it. If this is something you think you might be interested in, agree on the price before you sign the lease.

(6) At the end of the lease agreement, you don't have to worry about the hassle of selling the car.

The Disadvantages Are:

(1) You have to sign up for a set amount of time. If, for some reason, you need to get rid of the car because you have to move, are going traveling, or change jobs, for example, you are bound to the contract and it can be very expensive to get out of.

(2) You are limited to a strict amount of miles you can drive every year (typically fifteen thousand miles); if you go over this amount, you have to pay for each additional mile you drive, and this can be pricey.

(3) Alternatively, if you drive only a fraction of the agreed-upon mileage, then you are paying for miles that you never used.

(4) When you are finished with the car, you have nothing to show for it. You have no car equity.

(5) You don't actually own the vehicle, so you can't make any modifications to it.

(6) You have to keep the car in good condition. There is obviously some unavoidable wear and tear that comes with driving, so a lease contract will typically have a "wear and tear" agreement, but if you are a fender-bender driver or need to transport kids or pets, you might cause more damage than is expected, and you will have to pay for it in the end.

The Ideal Candidate

❶ You drive around fifteen thousand miles a year.

❷ You keep your car in good condition.

❸ There are no future circumstances—that you are aware of, anyway—that will force you to sell your car in the next three years.

If you decide that leasing is for you, the most important thing is to know you trust the company you are dealing with. Read the agreement very carefully and make sure you are happy with all the clauses. If you are unsure of what something means, don't be afraid to ask. Remember, the person you are dealing with is a sales rep on commission. No matter how aloof they may appear to be, they want to get your business, so don't be afraid to haggle or get them to change parts of the agreement that aren't suited to you.

BUYING A USED CAR

Advantages

The main advantage of buying a used car is that you get great value for your money. Fifteen thousand dollars might buy you a new economy-size car or a five-year-old sporty little number.

If it is up to three years old, the car could still be under the manufacturer's warranty, which you can still take advantage of should you have any problems.

While there is a risk involved in buying a used car (see the "disadvantages" section below), there are now lots of car-history-check companies out there that can help minimize the risk.

Disadvantages

The main disadvantage when you are buying a used car is the risk involved. There are lots of things that can happen to a car during its lifetime that can greatly devalue it (such as being involved in a big crash or having lots of engine trouble), and the person selling it to you might not disclose this properly, in order to get as much money from

you as possible. Also, there is very little legal protection when buying a used car, particularly in a private sale.

But have no fear; there are ways to minimize the risk. Yes, there is slightly more hassle involved, but if you know what you are doing, you can get a great deal.

TRICKS THE BAD GUYS PLAY AND HOW TO SPOT THEM

Unfortunately, there are some bad guys out there who make a lot of money by obtaining cars with a shady background and then selling them for a lot more money than they are worth. You could potentially fork over all your money only to find out shortly afterward that all is not what it  seems, and you may even have to hand the car over to the police or previous owners (in the event that it was stolen) without getting one penny back. You may as well have taken your money and flushed it down the toilet. So, outlined below are the tricks people try to play, but fortunately, there are ways to spot them!

Unpaid Financing

Outstanding unpaid financing on a car is one of the most common problems that happens today. If someone buys a car with a loan, technically the car is the property of the bank, or institution that the borrowed money came from, until all the monies have been paid. However, some people try to sell their car while they still have an outstanding loan on it. If you buy a car that has an outstanding loan on it, when you go to get insurance, the banks will track that a car they technically still own is being sold off, and they have the right to repossess the car from you. The best way to avoid this, and most of the other common problems, is to run a background check on the car. Go to the end of this section for more details.

Clocked

If a car has been clocked, it means that someone has tampered with the odometer—the device on the dashboard that says how many miles the car has driven in total. These culprits usually reduce the figure to make it appear that the car has done fewer miles than it actually has so you will pay more for it.

Again, a background check can sometimes pick up whether a car has been clocked. Alternatively, a good idea is to look at the service manual. If the last service stamp says it was carried out at 60,000 miles and the odometer reads 40,000, you know it has been clocked.

Salvage Title

If a car has been involved in a serious accident, an insurance company can sometimes declare that the cost of repairing the car is much more than it's worth. Unfortunately, some shady mechanics can fix up these cars so that they look okay and then go on to sell them without disclosing the car's history. Your background check should tell you if the car has ever been salvaged.

Stolen

Thousands of cars get stolen every year with the objective of being sold again for profit. If you buy a stolen car, when you go to do things like pay for the registration and insurance, you will have to provide your license plate number, and as these companies have databases of stolen license plates, alarm bells will start ringing. You will get a nice, friendly visit from a police officer who will take the car from you, and no matter how innocent you are, you won't see a penny.

Cloned

Car thieves can be clever little chaps and are becoming very good at disguising the fact that a car was stolen. One of their tricks is called cloning. Cars are usually identified by their make, model, and license plate. If theives steal, say, a red 2003 Honda Civic, they will look for a similar red 2003 Civic on the road and take down its license plate details, get a copy made, take the plates off the stolen car, and put the copied plates on it. Now there are two very similar cars driving around. It is more than likely that the owner of the original car has no idea their car has been cloned. The cloned car can pick up speeding tickets, and parking fines or, worse, be involved in a hit-and-run or robbery, but they will all be traced back to the innocent person's home address!

Ringing

Ringing is another trick to disguise a stolen car. Thieves steal a car and find a similar car that was totaled, steal its license plates, and put them on the stolen car.

Cut 'n' Shut

Say a 2004 Opel Tigra was involved in a crash and the front was completely destroyed. Thieves will find another 2004 Opel Tigra that was involved in a crash where the back was completely destroyed. They will then weld together the "good" parts of both the vehicles to make one car.

A good welder can do a superb job of hiding the seams, but if you buy the car and are involved in an accident, the joints can be very weak and will crumple under the pressure of the crash. Again, a car-history check should determine this for you.

How Can I Prevent This?

While the license plates can be changed easily, the vehicle identification number (VIN) cannot. All cars have to be assigned a unique VIN when they are manufactured in the factory. They are typically made up of numbers and letters and are seventeen digits long.

Where Can I Find My VIN?

Look up "Vehicle Identification Number" or "Chassis Number" in the owner's manual of the car you are interested in. That should tell you where it is located. The VIN can be put in different places on different cars, but here are a few common places to look. When you have found the VIN, give this to the car-history-check company, and they will be able to tell you if it has been stolen.

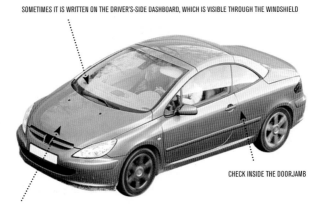

SOMETIMES IT IS WRITTEN ON THE DRIVER'S-SIDE DASHBOARD, WHICH IS VISIBLE THROUGH THE WINDSHIELD

CHECK INSIDE THE DOORJAMB

RAISE HOOD AND SEE IF IT IS ON THE FRONT OF THE ENGINE

Finding Out the History of Your Car

The best way to find out the history of your car is to use a car background check. These are relatively inexpensive but are worth their weight in gold. Before you even go to the effort of checking the car, ask the seller for the license plate number and VIN of the car, and get it checked out.

Companies that carry out such checks include:

- www.carfax.com
- www.autocheck.com
- www.carproof.com (popular in Canada)

Going to Look at a Secondhand Car

Okay. So, you've found the car of your dreams, you've done the math, and you've figured out that you will be able to afford it. Now you have to go and look at it—but what, exactly, should you look at? Ideally, you want a car that has been well maintained and doesn't appear to have that much wrong with it. What you don't want is to buy a used car with loads of problems that ends up becoming a money pit for you. Going to have a proper look at the car is essential to make sure you don't have any major problems after you buy it.

Here is a list of the things you need to check. None of these is hard—it just might take you a while to carry it all out. But you are about to spend a lot of money, so you want to make sure you are making the right decision.

Don't assume that if you bring a guy with you he will know what he's doing. Bring this book with you and run through it as you go along, if need be.

(1) Arrange to look at the car in the morning, so that if it has been parked in the same place overnight, you will be able to see if it has leaked any fluid.

(2) When you first meet the seller, ask them to give you a history of the car: Has it been involved in any accidents? Have there been any major problems with the engine? They may disclose these things to you and may be offering you the car at a reduced price. But if they don't tell you and you find things wrong, then they are hiding stuff from you, and your alarm bells should start ringing.

(3) Ask them for a copy of the service history and the owner's manual. Look at the service history. It should be dated and stamped by every mechanic who did work on it. A car should be serviced every six months to keep it in top order. Have they been doing this? Does the service manual say that they got any major work done that they didn't tell you about? They may say they have lost the book. It does happen; if so, ask them for the name of their mechanic, call him up, and ask him directly how often he tended to the car and what he did.

(4) Look up "Timing Belt" in the owner's manual. It should say when the belt needs to be changed. How many miles has the car done? Is it past the mark, or is it coming close to when the timing belt needs to be changed? If so, ask the person about it. They got it changed? If so, ask for proof. If not, this is something that you will have to pay for soon. Just be aware that it is going to be an added expense in the near future.

Under the Hood

⊛ Open the hood. Ask the seller to do it for you, or look under "Hood" in the service manual and it will tell you how to open it.

⊛ Check the oil, coolant, and brake fluid to make sure they are all okay. Refer to Chapter 3 to find out how to do this.

⊛ Now have a look at the engine and inspect any belts and hoses you can see. Do they look cracked or worn?

⊛ Look around the radiator. If there are white stains around it, it could be a sign that the engine has overheated at some point.

⊛ Have a look at the battery and make sure it looks clean enough. Ask the seller when the last time they got it changed was. Batteries need to be changed roughly every three years. If it has been a while since they got it changed, you will have to shell out money for a new one soon.

Walking Around the Car

Now that you have checked out the engine, you need to look at the external things.

Under the Car

Get down on your knees and look under the car. Can you see any patches that could indicate a leak?

Tires

Have a careful look at the tires and make sure they are all in good condition. Make sure there are no cracks or bumps and that there's no excessive wearing. Make sure the tread depth is okay on the tires and that they are not bald. Check them all, and don't forget the spare. If they look old and worn, remember it is illegal to drive with poor tires, so you will have to pay for new ones.

Suspension

Walk around to each corner of the car and give it a good push down. The car should bounce back once or twice and then stop. This is a good indication that the suspension is okay.

Body of the Car

Walk around the car and carefully look at the bodywork. Look out for any noticeable scrapes, dents, or rust. Check all of the seams, too (where two pieces of metal meet, e.g., the door frames or hood), and look for any signs that this section of the car has been resprayed. This could be an indication that it has been involved in an accident.

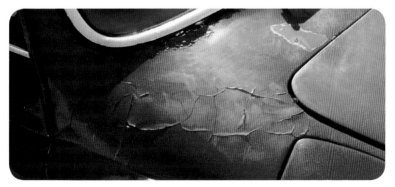

Interior

Now open all the doors, including the trunk, have a look at the upholstery, carpet, and vinyl, and make sure you are happy with the appear-
ance. These features don't actually affect the car's driving ability and
safety, but if they are excessively worn, you can use that point as a good
bargaining tool for reducing the price.

Behind the Driver's Seat

Leaving the car in park, turn on the engine.

- Open and close all the windows.
- Turn on the radio to make sure it is working.
- If the car has air-conditioning, test it.
- Turn on the wipers.
- Test the locks.
- Now turn on all the lights, including the hazard lights, get out of the car,
  walk around all four corners, and make sure the lights are all working.

THE TEST-DRIVE

Now that you have confirmed that you are happy with everything, you will need to take the car for a test-drive. If you already own a car, make sure your own insurance policy covers you when you drive someone else's car.

If you are a first-time driver and don't have insurance yet, you will need to get someone else who is covered to drive it for you. Sit in the front seat with them and make sure they carry out the following checks.

The Engine

As you are driving, turn off the radio and air-conditioning, open the windows, and listen to the sound of the engine. It should be smooth and should not make too much noise. Then turn your head toward the back of the car and listen to the exhaust. It should be reasonably quiet also.

Suspension

If you can, try to drive the car over some speed bumps, and make sure you are happy with the suspension.

The Brakes

Test the brakes in normal traffic and against traffic lights. Make sure they work efficiently; there should be no noises coming from them.

When it is safe to do so, try to get the car to stop suddenly. How did the brakes react?

Test the hand brake and make sure it pulls up easily.

The Gears

The gear change should be smooth and easy. If it is very difficult to change gears, then it could mean that the gearbox is nearing the end of its life. You will have to replace it sometime in the near future, and that will be very expensive.

Wheel Alignment

When you are driving on a straight stretch of road, really loosen your grip on the steering wheel. The car should continue to drive straight for you. If it veers off to one side, this means the wheel alignment is off. This can be easily fixed—for a fee, of course.

Convertible

If you are buying a convertible, make sure you test the top—go through the whole process—and make sure there are no problems.

Professional Check

So, if you are happy that there is nothing majorly wrong with the car, you can still get a mechanic to check the car properly for you. You are going to have to pay him for this, but it might be worth it for peace of mind.

HOW MUCH SHOULD I PAY FOR IT?

You have found the car of your dreams, have done a history check, have inspected it yourself, and are happy with it. How much should you actually be paying for it?

Go online and look at www.whatcar.com for a guide price. If you have gotten a mechanic to check it out, ask him how much he thinks it is worth. If you haven't gone the mechanic route, look online and see how much cars similar in age and mileage are selling for. This will give you a starting point, but you should be able to haggle down the price if any of the following things are wrong:

* Your car has excessive mileage compared with other cars for sale that are the same year.

* There are scrapes and scratches on the body. These ultimately won't affect the driving ability or safety of the car, so it is a personal choice whether you mind driving with them, but you should definitely be looking for a few hundred off the asking price if they are there.

* The interior hasn't been well maintained.

If there are problems that can be fixed—such as that the car needs new tires or wheel alignment—you can take away the cost of repairing these from the quote, but remember, you will end up with the hassle and expense of doing it yourself.

The Paperwork

Once you have handed over all the money for the car, you need to fill in your details on the vehicle registration certificate and the former owner needs to sign it. The certificate then needs to be delivered to the DMV in order for the car to be legally registered as yours.

Also ensure you get the owner's manual, service log, and spare keys from the existing owners, and that's it! Congratulations on getting a new car!

Chapter Nine

SELLING A CAR

WHEN IT COMES TO SELLING

a car, you have three options.

(1) Sell it to a car dealer.

(2) Get the price of the car offset against buying a new car from a dealer—known as a "trade-in."

(3) Sell the car privately yourself.

If you sell your car to a dealer, you can avoid the hassle of selling a car, but as the dealer still needs to sell it and make a profit, you will get the lowest price possible for the car. If your car is in bad condition, though, sometimes dealers can offer the best price, as they have the means of fixing your car cheaply.

HOW MUCH IS IT WORTH?

The quickest way to find out how much your car is worth is to go online.

Websites such as Kelly Blue Book, www.kbb.com, are great for price quotes and comparisons for buying or selling a car.

You can actually figure out this information for free. Go onto websites that sell cars, such as www.autotrader.com, and look at how much cars that are the same make, model, and year as your car, with similar mileage, are selling for. Write down three or four prices, and calculate an average based on these prices.

IF GOING ONLINE JUST ISN'T FOR YOU, THEN YOU CAN PURCHASE A CAR-SELLING MAGAZINE TO GET THE AVERAGE PRICE.

This price will give you a starting figure to work from. If your car has lots of extra features or is a special edition, then you can look to sell it for the price of a car that is a year younger than yours.

WHAT CONDITION IS YOUR CAR IN?

Next, you need to determine what condition your car is in. Have a good look at it and determine the following:

ENGINE	EXCELLENT	EXPECTED FOR AGE	POOR
HAVE YOU BEEN HAVING ANY PROBLEMS WITH YOUR ENGINE LATELY? HAVE YOU BEEN GOOD AT TAKING YOUR CAR FOR REGULAR SERVICING AND KEEPING A RECORD IN YOUR MAINTENANCE LOGBOOK?			
PAINTWORK			
ARE THERE ANY SCRATCHES OR DINGS ON YOUR PAINTWORK, OR IS IT AS GOOD AS NEW?			
TIRES			
HOW DEEP IS THE TREAD IN YOUR TIRES? ARE THERE ANY CRACKS OR BULGES?			
INTERIOR			
WHAT IS THE CONDITION OF YOUR UPHOLSTERY, CARPETS, AND VINYL? DON'T FORGET TO LOOK AT THE TRUNK.			
ELECTRONICS			
ARE YOUR RADIO, ELECTRIC WINDOWS, LIGHTS, AND AIR-CONDITIONING (IF YOU HAVE IT) WORKING OKAY?			

If your car is in excellent condition, you can expect to charge the top price for it. If it is in poor condition, you can expect the buyer to try to haggle down the asking price.

Once you have determined your price, you should add a few extra hundred dollars, as most people will expect you to negotiate. This will give you some extra leeway. Don't, however, set the price too high, as this will scare people off. If you are looking at prices online, you can expect that they include "negotiation" leverage, so perhaps follow their lead, but expect to sell your own car for a few hundred less.

Should I Fix It?

If you have something wrong with your car, such as engine trouble or outstanding necessary bodywork, you will have to decide whether you want to pay for it to be fixed or whether you would prefer to deduct the cost of the repairs from the asking price and get the new owner to pay for them. If you don't have the money to pay for the work yourself, don't worry—you would be surprised at how many people out there will take a car that needs work done to it for a reduced price. Just be sure to include the information on the ad.

PUTTING THE CAR UP FOR SALE

Currently, the most common place for people to look when buying a car is online. There are lots of websites, such as www.autotrader.com and www.craigslist.org, where you can place your ad. There are also car magazines that you can contact.

The Ad

Ads with pictures sell cars more quickly, so go out and clean the car, park it in a nice place, and take some good pictures of it.

Include the make, model, year, engine size, mileage, and color, as well as fuel type (gasoline or diesel) and transmission (manual or automatic).

Bring out the saleswoman in you, and note any special features your car has—but be honest.

Which of the following is true of your car? If you have it, include it!

⊛ MOT/NCT

⊛ Excellent condition

⊛ Good condition

⊛ Low mileage

⊛ Full service history

⊛ Power steering

⊛ New tires

⊛ Electric windows

⊛ Air-conditioning

⊛ Sunroof

⊛ Convertible

⊛ Color-coded bumpers
(when the bumpers are the same color as the rest of the car)

⊛ Airbags

⊛ Alloy wheels

⊛ CD player

⊛ Any other custom modifications

Viewing the Car

If someone is interested in buying your car, they will want to come inspect it. For safety reasons, it is better if they come to your house. Have someone there with you if you are worried about your personal safety. A first impression is everything, so make sure you clean the car just before they come. Have your owner's manual and maintenance log at the ready.

Test-Drive

The buyer will more than likely want to take the car for a test-drive. You are not obligated to allow them to take one, but it will probably increase the chances of sale if you do. Make sure that your car is insured to allow someone else to drive it or that the buyer is covered under their own policy. Unless you want someone to drive off with a free car, always go on the test-drive with them. Don't forget to fully lock your house before you leave.

Negotiating the Price

Know in your mind exactly what the minimum price is that you would be willing to sell your car for. Do not tell prospective buyers what your lowest price is, and do not go below it. If you have had several callers, this means you shouldn't need to haggle down on price. Some people are great at haggling. They will try to make you feel that your car is worthless and that you would be lucky to get their business at a greatly reduced price, but don't fall for their tricks. If they are offering you less than you feel is right, based on your market research, just say no. You might be surprised that they will suddenly pay your asking price. There is an art to haggling!

The Money Bit

The cardinal rule you need to remember here is: Do not hand over the car to anyone until the money is sitting in your bank account. Checks can bounce and banker's drafts can be forged, so make sure the money actually goes through before you release your car.

As cars are an expensive commodity, buyers rarely have enough cash on hand to pay the full amount immediately. It is more likely that they will want to put a deposit on the car and arrange financing. If someone is interested, make sure you get a nonrefundable deposit for a set amount of time. You do not want to be waiting around for a month to accept payment.

In the interest of both parties, it is important that you draft a proper receipt. Ensure that you make two copies—one for you and one for them. Here is a sample of what you need to include:

Seller's Details

- ✳ NAME _____
- ✳ ADDRESS _____
- ✳ CONTACT _____
- ✳ PRICE AGREED UPON _____
- ✳ DEPOSIT AMOUNT _____
- ✳ FINAL DATE FULL AMOUNT IS DUE _____
- ✳ SIGNATURE _____
- ✳ DATE _____

Buyer's Details

- ✳ NAME _____
- ✳ ADDRESS _____
- ✳ CONTACT _____
- ✳ PRICE AGREED UPON _____
- ✳ DEPOSIT AMOUNT _____
- ✳ FINAL DATE FULL AMOUNT IS DUE _____
- ✳ SIGNATURE _____
- ✳ DATE _____

Saying Farewell

Once the money has been paid and both parties are satisfied with the deal, you will need to hand over all your car documentation and spare keys to the new owner. The new owner needs to fill in their details on the car's registration documents, and you need to sign the

bottom of it. This paperwork needs to be delivered to the DMV.

THANK-YOUS

Jamie's thanks

Thank you to my mom for being my role model—the strong, independent woman I've always strived to be. Thank you to Rich Feinberg at ESPN for having faith in me from the beginning and allowing me to live my dream job. To my husband, Cody, for being my rock, my balance, and my love. And to my dad: Your strength, intelligence, and love have always been and are the best examples a child could ask for. And to my sis, Lindsay: Our bond from the beginning and your loving energy are a blessing.

Danielle's thanks

Writing a book is an interesting journey in which you meet amazing people along the way. First off, I'd like to thank my family, in particular my father, Tom McCormick. This book brought us closer together.

It is no small feat to find someone who has a deep understanding of engines and can also translate British to American English. But when it came to finding this person, I didn't have to look much further than one of my dearest friends, Rod Schenker, whose passion for engines and humanity led him to become a respected pilot and captain in the Marines. Luckily, his other passions of literature, existentialism, and beer led him to become friends with an Irish girl. I'm truly grateful for all the amazing friends I have in my life, who are always there to shout words of encouragement from the sidelines.

I'd also like to thank Michael Voorhees for taking such wonderful pictures. He is a gifted photographer and an inspiration, a person who followed his dreams and now leads an amazing and adventurous life. Thank you to Christina Copeland for making us girls look pretty and to Matt Campbell for helping out on the day of the shoot and coming to Jamie's rescue. Also, a big thank-you to Las Vegas Speedway for letting us shoot at their facilities.

Finally, I'd like to thank my agent, Jill Marsal, for believing in the project, and the team at Seal for making this American edition a reality.

About the Authors

JAMIE LITTLE is a well-known motor-sports reporter for ESPN and ABC. She covers ESPN's X Games, IndyCar, and Indy 500 and the ever-popular NASCAR Sprint Cup and Nationwide series. Little graduated from San Diego State University in 2001 with a degree in journalism. Born in Lake Tahoe, California, she's had a fascination with motorcycles from a young age. That interest turned into a love for motocross and eventually  auto racing. She loves everything with a motor. Little is a self-proclaimed tomboy who is an independent woman in a man's world. There's no task she won't tackle. That is one of the many reasons she feels this book is a must-have for women and girls everywhere. Little lives in Las Vegas, Nevada, with her husband, Cody, and their two dogs.

DANIELLE MCCORMICK hails from Ireland, where she is known for being the bestselling author of the original UK edition of *Essential Car Care for Girls*. The inspiration for the book came when she blew a head gasket on her car and was shocked to discover so few books for girls about cars. She was the only female car representative on Discovery Turbo's blog, and she now works with electric-car companies  to promote more sustainable driving. Danielle resides in Silicon Valley, where she is the CEO of a start-up company. In her spare time, she is an avid skier and the worst crew member of a very good yacht racing team.

INDEX

Selected Titles from Seal Press

By women. For women.

MARIE'S HOME IMPROVEMENT GUIDE, by Marie L. Leonard. $16.95, 978-1-58005-292-4. A practical how-to guide for women with to-do lists, Marie's Home Improvement Guide offers all the tips you need to tackle home repair projects . . . yourself!

SHE-SMOKE: BBQ BASICS FOR WOMEN, by Julie Reinhardt. $14.95, 1-58005-284-3. The owner of Smokin' Pete's BBQ in Seattle lays down all the delicious facts for women who aspire to be BBQ queens.

THE QUARTER-ACRE FARM: HOW I KEPT THE PATIO, LOST THE LAWN, AND FED MY FAMILY FOR A YEAR, by Spring Warren. $16.95, 978-1-58005-340-2. Spring Warren's warm, witty, beautifully illustrated account of deciding—despite all resistance—to get her hands dirty, create a garden in her suburban yard, and grow 75 percent of all the food her family consumed for one year.

THE BOSS OF YOU: EVERYTHING A WOMAN NEEDS TO KNOW TO START, RUN, AND MAINTAIN HER OWN BUSINESS, by Emira Mears & Lauren Bacon. $15.95, 1-58005-236-3. Provides women entrepreneurs the advice, guidance, and straightforward how-to's they need to start, run, and maintain a business.

RUN LIKE A GIRL: HOW STRONG WOMEN MAKE HAPPY LIVES, by Mina Samuels. $16.95, 978-1-58005-345-7. Author and athlete Mina Samuels writes about how lessons learned on the field (or track, or slopes) can help us face challenges in other areas—and how, for many women, participating in sports translates into leading a happier, more fulfilling life.

UNDECIDED: HOW TO DITCH THE ENDLESS QUEST FOR PERFECT AND FIND THE CAREER—AND LIFE—THAT'S RIGHT FOR YOU, by Barbara Kelley and Shannon Kelley. $16.95, 978-1-58005-341-9. Mother and daughter Barbara and Shannon Kelley explore how women's choices have evolved, why it's so overwhelming, and what we can do about it—starting with a serious shift in perspective.

www.SealPress.com
www.Facebook.com/SealPress
Twitter: @SealPress